Global Environmental Change and International Relations

Edited by

Ian H. Rowlands

Department of Internatio

London School of Econon ... olitical Science

and

Malory Greene

Department of International Relations

London School of Economics and Political Science

M in association with *Millennium: Journal of International Studies*

First published 1992

Published by
MACMILLAN ACADEMIC AND PROFESSIONAL LTD
Houndmills, Basingstoke, Hampshire RG21 2XS
and London
Companies and representatives
throughout the world

Edited and typeset by Grahame & Grahame Editorial, Brighton

ISBN 0–333–55490–6 (hardcover)
ISBN 0–333–55491–4 (paperback)

A catalogue record for this book is
available from the British Library

Printed in Hong Kong

To

Cameron and Stephen

I.H.R.

Theresa and Jeremy

M.G.

'We have not inherited
the earth from our parents –
we have only borrowed
it from our children.'

Contents

List of Tables and Figures

Tables

Figure

Acknowledgements

This book has its origins in a Special Issue of *Millennium: Journal of International Studies* (Volume 19, Number 3, Winter 1990), which is published from the Department of International Relations at the London School of Economics and Political Science. Ian conceived the idea of a Special Issue on the subject in 1989 and brought it closer to fruition by organising a panel on 'The International Relations of Global Warming' for the annual conference of the International Studies Association in Washington, DC in April 1990. The Special Issue was only realised after the efforts of several people.

The entire management board of Volume 19 (1990) of *Millennium* contributed to the production of the Special Issue: Book Reviews Editors Richard Jerram and Janet LeNoble; Deputy Editors John Hannaford and Sophia Okpala; Associate Editors Kathleen Newland, Peter Wilson and the late R. J. Vincent; and Business Manager Priya Mukherjee. We are particularly grateful to Kathleen for her constant support, encouragement and willingness to assist us in whatever way we needed. Especially under hectic circumstances, she provided a much needed calming influence. Other individuals previously associated with the Journal also contributed. Rebecca Grant, Mark Hoffman and David Long encouraged us to take on the editorship of the journal and undertake the Special Issue, and Hugh Dyer and Spyros Economides were always on hand to help out during the production process. We are also grateful to the institutional support provided to *Millennium* by the London School of Economics, in particular its Department of International Relations. Michael Banks' support of our endeavour is also appreciated. The authors in this book deserve special thanks for producing their chapters on a very tight schedule – at times communicating by fax, electronic mail and telephone to ensure that deadlines were met.

We would also like to thank our publishers, who encouraged us to turn the Special Issue of *Millennium* into a book. Thanks also to MIT Press and the Center for Strategic and International Studies in Washington, DC for permission to print a different version of Ian's article, which originally appeared in *The Washington Quarterly*. Finally, we are grateful for the constant support that our families have provided us – not only in the production of this book, but with all our academic endeavours.

IAN H. ROWLANDS and MALORY GREENE

Notes on the Contributors

Daniel Deudney is Hewlett Fellow in Science, Technology and Politics at the Center for Energy and Environmental Studies at Princeton University, Princeton, NJ. He received a B.A. in Political Science and Philosophy from Yale University, a Masters in Science, Technology and Public Policy from George Washington University and a Ph.D. in International Relations from Princeton University. He was previously a Senior Legislative Assistant on Energy and Environment for a US Senator and a Senior Researcher at the Worldwatch Institute. He is co-author of *Renewable Energy.* His present research interests include global geopolitics and he is writing a book entitled *Pax Atomica: States and Republics in Sustainable Global Security Systems.*

Malory Greene is a Ph.D. candidate in the Department of International Relations at the London School of Economics and Political Science. She received a B.A. in Political Science from the University of California, Los Angeles and an M.A. in Latin American Studies from the London School of Economics and Political Science. She is writing her dissertation on the political factors affecting Mexican trade policies. She was previously editor of *Millennium: Journal of International Studies.*

Peter M. Haas is Assistant Professor in Political Science at the University of Massachusetts at Amherst, MA. He has also taught at Yale University and has been a research fellow at the Woods Hole Oceanographic Institution's Marine Policy Center. He holds a Ph.D. in Political Science from the Massachusetts Institute of Technology. He is author of *Saving the Mediterranean: The Politics of International Environmental Cooperation* as well as a number of articles and chapters dealing with international environmental issues. His research interests include international environmental politics, the role of science in decision-making and economic development.

David Pearce is Professor in Economics at University College London. He is also Associate Director of the London Environmental Economics Centre and Director-Designate of the Centre for Social and Economic Research on the Global Environment. He has held prior positions at Aberdeen University, Leicester University, the University of Southampton and Lancaster University. He is the author of over 150 articles and author, co-author or editor of 25 books including *Blueprint for a Green Economy, Economics of Natural Resources and the Environment, Sustainable Development: Economics and Environment in the Third World* and *Cost-Benefit Analysis*. He is advisor to the UK Secretary of State for the Environment and is a recipient of the United Nations Environment Programme's *Global 500* roll of honour for services to the world environment.

Glen Plant is Lecturer in International Law at the London School of Economics and Political Science. He is also a practising barrister. He graduated in Jurisprudence from New College, Oxford, received his Ph.D. from King's College, London, and obtained the Diploma in International Law from the Hague Academy of International Law. He has held positions at King's College, London, Durham University and the Fletcher School of Law and Diplomacy, Tufts University. He has also served as a Law of the Sea Officer for the United Nations in New York and Jamaica, and served as an Assistant Legal Adviser in the Foreign and Commonwealth Office.

Kilaparti Ramakrishna is Senior Associate for International Environmental Law at the Woods Hole Research Center. He obtained a Ph.D. in International Environmental Law from Jawaharlal Nehru University, New Delhi. He worked as Assistant Professor at the Indian Academy of International Law and Diplomacy and at the Indian Society of International Law prior to attending the Harvard Law School as a Fulbright Visiting Scholar. He has held teaching and research positions at the Woods Hole Oceanographic Institution, the East-West Center and Boston University. In addition to global warming and climate change issues, his research interests include national and international decision-making trends, alternative dispute resolution mechanisms and the management of marine sanctuaries.

Ian H. Rowlands is a Ph.D. candidate in the Department of International Relations at the London School of Economics and Political Science. He received a B.A.Sc. in Engineering Science from the University of Toronto and a Diploma in International and Comparative Politics from the London School of Economics and Political Science. He is writing his dissertation on the factors that affect international co-operation on issues of global environmental change. He was previously editor of *Millennium: Journal of International Studies*.

Oran R. Young is Research Professor of Government, Senior Fellow of the Dickey Endowment for International Understanding and Director of the Institute of Arctic Studies at Dartmouth College, Hanover, NH. He also serves as chair of the Committee on the Human Dimensions of Global Change of the National Academy of Sciences. He has previously held faculty appointments at Princeton University, the University of Texas at Austin, the University of Maryland at College Park and the Center for Northern Studies. An authority on collective decision-making relating to natural resources and the environment and on the Arctic, his most recent books are *International Cooperation: Building Regimes for Natural Resources and the Environment* and (with Gail Osherenko) *The Age of the Arctic: Hot Conflicts and Cold Realities*.

Craig ZumBrunnen is Associate Professor of Geography at the University of Washington, Seattle, WA. He also holds faculty appointments in the Russian and East European Unit of the Jackson School of International Studies and the Institute of Environmental Studies at the University of Washington. Previously, he was Assistant Professor of Geography at Ohio State University. He received a B.A. in Geology from the University of Minnesota, an M.S. in Geological Sciences from the California Institute of Technology and a Ph.D. in Geography from the University of California, Berkeley. He is the co-author of *The Soviet Iron and Steel Industry*, editor of *Urban Geography in the Soviet Union and the United States* and author of over 40 articles and book chapters dealing primarily with Soviet environmental and resource issues.

1 Introduction

Ian H. Rowlands and Malory Greene

The issues of global environmental change, as epitomised by the problems of ozone layer depletion and global warming, are firmly entrenched upon the international agenda. During the late 1980s and the early 1990s, the consequences of human intervention in the earth's natural systems became evident in two significant ways. First, scientific investigations after the 1985 discovery of an ozone 'crater' above Antarctica revealed that chlorofluorocarbons and other widely-used substances were destroying the earth's protective ozone layer. And second, in 1990, the Intergovernmental Panel on Climate Change concluded with certainty that

> emissions resulting from human activities are substantially increasing the atmospheric concentrations of the greenhouse gases . . . These increases will enhance the greenhouse effect, resulting on average in an additional warming of the Earth's surface.[1]

Such changes in the earth's natural equilibria could have a profound impact upon international society.

Environmental issues in international relations are, of course, not new. It has long been acknowledged that the possession, distribution and exploitation of strategic natural resources are a key determinant of the rise and fall of states. Additionally, geopolitical considerations have been part of the foreign policy calculations of states' decision-makers since the beginning of the modern state-system. Indeed, both Jean Bodin and Montesquieu noted the influence of climatic factors upon political activity. Bodin 'maintained that climatic circumstances influence national characteristics, even determining their foreign policies', while Montesquieu 'pointed to various climatic factors which he felt influenced man's political behavior and vitality'.[2] In the nineteenth century, concern about the management of natural resources can be viewed as one of the factors contributing to the emergence of international organisations. International commissions to regulate traffic and trade along various European rivers were established – for example, the Elbe in 1821, the Rhine in 1831 and the Danube in 1856.[3] Formal international discussions about transboundary pollution date back to at least the 1930s. In 1935, an international tribunal concluded that noxious fumes produced by a Canadian smelter had been travelling down the Columbia River valley, crossing the Canada-US border and damaging American interests in the state of Washington.[4]

More recently, a number of significant events in the 1960s intensified national concerns about environmental issues and thus helped to propel the general question up the international agenda. The publication of Rachel Carson's *Silent Spring*,[5] the crash of *The Torrey Canyon* and the debate concerning overpopulation and resource depletion are but three examples. Oil and food crises during the 1970s, combined with the considerable impact of the Club of Rome's *Limits to Growth* report,[6] kept environmental issues as a topic of international discussion.

In the past, however, most environmental issues were thought to be only local, national or, at most, international concerns narrowly defined – that is, concerns which affected two or more, but nevertheless a limited number of states. It was generally assumed that wrongs could be immediately righted and that any environmental degradation could be contained over both space and time.

Owing to the challenges of global environmental change, however, this way of thinking must now be reconsidered. The new issues are not merely 'international' but are truly 'global', because no state is immune from the worldwide changes. Further, given the systemic nature of the issues, there are considerable lagtimes involved. Thus, preventative and anticipatory action and the question of intergenerational fairness are of far greater importance.

The emerging challenges of global environmental change are having, and will continue to have, implications for the way that international relations is both practiced and studied. With regard to its practice, all actors in the international arena are beginning to recognise and acknowledge the links that exist between the natural earth and the anthropocentric world. Decision-makers are, therefore, now compelled to take into account the impact that proposed actions may have on the environment – and vice versa. This interchange is transforming the practice of international relations. Meanwhile, within the academic discipline, it is increasingly being acknowledged that the dominant interpretations, concepts and frameworks are being challenged by the new global environmental issues. It is becoming clear that a variety of approaches and methodologies must be employed in order to gain a fuller understanding of the way in which global environmental change affects world politics.

The contributions in this volume investigate the implications for international relations of global environmental change, exploring both policy and theoretical problems. The variety of perspectives employed – international relations theory, international political economy, Eastern Europe, international law, North-South issues and security studies – illustrates the range and complexity of the issues involved. The investigation is far from exhaustive – but it is our hope that this book will open new areas of discussion in an increasingly vigorous debate.

In Chapter 2, Oran Young considers the problems of global environmental change in a broad theoretical discussion of international relations. He identifies five themes of research that have emerged in the study of international environmental regimes: structural, power-based, interest-based, cognitive and contextual. Young asserts that the range of inquiries should be broadened so as to include investigations into preparatory activities leading up to institutional bargaining, the problems associated with North-South interactions, the role of non-state actors and the relationships between institutions and organisations.

Ian Rowlands provides a historical overview of the ozone layer depletion and global warming issues in Chapter 3. For each of the two issues, he provides a brief description of the scientific processes involved and recounts what are purported to be the most significant social system impacts of these natural changes. Rowlands also reviews the history of international political activity in these areas and provides some preliminary thoughts about the major issues that may lie ahead for international society.

Peter Haas, in Chapter 4, examines the role that 'epistemic communities' can play in achieving international co-operative agreements on environmental issues. Epistemic communities are transnational networks of knowledge based communities that are both politically empowered through their claims to exercise authoritative knowledge and motivated by shared causal and principled beliefs. Haas argues that, although an epistemic community helped to create the existing regime to protect the ozone layer, agreements are unlikely to be achieved by the same process in the global warming issue.

In Chapter 5, David Pearce looks at the issues from an international political economy perspective. He initially lays out a 'constant natural capital' framework and argues that international agreements on ozone layer protection fit into this structure. This framework is supplemented by a utilitarian approach in order to open discussion on global climatic change. Pearce subsequently examines three possible routes that can be taken to achieve internationally agreed global environmental targets: command-and-control measures, pollution taxes and tradeable permits. He argues that the latter two – both being market-based instruments – hold the most promise for meeting the challenges of global warming.

The unique environmental challenges in Eastern Europe are investigated by Craig ZumBrunnen in Chapter 6. With the waning of the Cold War and the dismantling of the iron curtain, the serious extent of the environmental degradation in this area has been revealed. ZumBrunnen examines the form and the magnitude of the environmental problems in the countries of this region. He considers how political and economic liberalisation in Eastern Europe may possibly affect the ability of the various states to deal with environmental degradation.

As the world community confronts global environmental change, decision-makers will have to decide whether to reform existing international institutions or to create new ones. Glen Plant, in Chapter 7, reviews the roles that the major institutions – both intergovernmental and nongovernmental – have played in the global warming debate thus far and considers what possible roles they may play in the future. Plant asserts that the realities of the international political system dictate that the most plausible scenario involves an enhanced status and role for the United Nations Environment Programme, along with its rationalisation within the UN system.

North-South issues are coming to the forefront of the global environmental debate, especially because the industrialised countries, which have been primarily responsible for much of the environmental degradation worldwide, are now asking the developing states to assist in the clean up. In Chapter 8, Kilaparti Ramakrishna argues that neither the Law of the Sea Convention nor the ozone layer agreements can be used as models for a law of the atmosphere. Further, the more general concept of the Common Heritage of Mankind also encounters difficulties when applied to the global climate change issue. Ramakrishna contends that developing a new approach which involves addressing issues of concern to developing countries, such as financial requirements, technology transfer and economic reforms, will help guarantee the developing countries' participation in the conclusion of meaningful conventions on global environmental change.

Daniel Deudney builds upon the emerging debate about the relationship between the environment and concepts of security in Chapter 9. In this chapter, he makes three assertions. First, Deudney argues that it is analytically misleading to think of environmental degradation stemming from global environmental change as a national security threat. Second, he contends that declaring global environmental change a threat to national security in order to mobilise action may prove counterproductive and produce undesirable side-effects. Third, Deudney asserts that global environmental change is not likely to cause interstate wars.

Taken together, these chapters represent an attempt to examine comprehensively the impact that global environmental change may have upon international relations. Evidence is increasingly pointing to the conclusion that the earth's natural systems are undergoing changes at unprecedented rates. There are still, however, differences of opinion about the manner, the magnitude and the distribution of the effects that such changes will have upon international society. Nevertheless, it is evident that the issues of global environmental change are no longer solely the concern of natural scientists; the issues are clearly also on the social scientists' agenda. Thus the 1990s is bound to be a pivotal decade with regard to the way in which society interacts with the natural earth. The decisions taken during this decade – resulting in either action or inaction – could well have the potential to transform the structure of international society. In this way, it is apparent that we are entering a critical time for practitioners and students of international relations.

Notes

1. WMO and UNEP, *The Policymakers' Summary of the Report of Working Group I to the Intergovernmental Panel on Climate Change* (Geneva: WMO and UNEP, 1990) p. 2.
2. Jean Bodin, *Six Books of the Commonwealth*, translated by E. J. Tooley (New York: Macmillan, 1955) pp. 145–57; and Baron de Montesquieu, *The Spirit of Laws* (Worcester: Isaiah Thomas, 1802) I, pp. 154–9 and 259–74. Both sources noted in James E. Dougherty and Robert L. Pfaltzgraff, Jr., *Contending Theories of International Relations* (New York: J.B. Lippincott Company, 1971) p. 46.
3. Clive Archer, *International Organizations* (London: George Allen & Unwin, 1983) p. 12.
4. Lynton Caldwell, *International Environmental Policy: Emergence and Dimensions* (Durham, NC: Duke University Press, 1984) p. 105.
5. Rachel Carson, *Silent Spring* (London: Penguin, 1965).
6. Donella H. Meadows, Dennis L. Meadows, Jorgen Randers and William W. Behrens III, *The Limits to Growth* (New York: Universe Books, 1972).

2 Global Environmental Change and International Governance

Oran R. Young

No knowledgeable observer anticipates the emergence of a world government during the foreseeable future; many would find such a development unappealing, even if it were feasible. Yet the growing salience of policy issues arising in connection with global environmental changes – like ozone layer depletion, global warming and the loss of biodiversity – has already stimulated a surge of interest in responses featuring agreements that call for sustained international co-operation. This has led in turn to a proliferation of efforts to form new international regimes or, more broadly, international institutions to prevent or control the impacts of global environmental changes.

The reasons for this development are easy to pinpoint. In many cases, efforts to cope with large-scale environmental changes cannot succeed in the absence of concerted action on the part of two or more states. Consider the issues of ozone layer depletion and global warming as cases in point. Recent initiatives on the part of the advanced industrial states to establish a timetable for phasing out the use of chlorofluorocarbons (CFCs) cannot solve the problem of ozone layer depletion unless some mechanism is devised to persuade China, India and other developing countries to use substitutes for CFCs in their rapidly increasing production and consumption of refrigerants. This is why the June 1990 amendment to the 1987 Montreal Protocol, calling for the establishment of an international fund to compensate developing countries for switching from CFCs to other, more benign technologies, is so important.[1] The case of global warming is even more complex. Not only is there a need for international co-operation to avoid a situation in which the efforts of advanced industrial states to reduce their carbon dioxide emissions are nullified by increased coal consumption on the part of China and other developing countries, but there is also the problem of co-operating to control other sources of greenhouse gases.

These sources range all the way from the release of methane in connection with rice production in South Asia through the unlocking of carbon stored in moist tropical forests to the use of CFCs in a variety of goods and services.

In other cases, global environmental changes stem largely from activities that take place within national jurisdictions, but, at the same time, generate spillover effects or externalities affecting the welfare of those residing elsewhere. Consider the destruction of moist tropical forests in the Amazon Basin and emissions of airborne pollutants in Eastern Europe as illustrative cases. In these instances, the need for international co-operation arises from the fact that the full costs of the actions in question are not factored into the decisions of those responsible for the actions. Under these conditions, there is a need to devise international mechanisms to articulate and apply liability rules covering the damages to others arising from emissions of carbon dioxide and other pollutants or, more likely, to allow those impacted by the spillover effects (typically advanced industrial countries) to compensate those responsible for the offensive emissions for terminating or redirecting their activities.

As these observations suggest, the growing interest in international co-operation to cope with global environmental changes is currently directed more towards preventing or mitigating anticipated impacts than towards adjusting to changes that have already occurred. The cases referred to in the preceding paragraphs, for example, involve such anticipatory actions as the phasing out of CFCs, reductions in the use of fossil fuels and the protection of moist tropical forests. As global environmental changes begin to become current realities, rather than future possibilities, however, the need for sustained international co-operation to adjust to the impacts of these changes will grow. If increasingly volatile weather patterns make agricultural production less predictable, for instance, many countries will discover growing incentives to establish international arrangements featuring buffer stocks and various types of agricultural insurance schemes. If the disruptive impacts of global environmental changes become even more severe, pressures will mount to devise international mechanisms to handle flows of environmental refugees crossing national frontiers, with or without the permission of recipient states. Today's preoccupation with setting up international arrangements to mitigate global environmental changes appears to be more a reflection of the nature of our current concerns than an indicator of the eventual role of international agreements in responding to large-scale environmental changes.

THE PROBLEMS OF CO-OPERATION

While the importance of international co-operation in coming to terms with global environmental changes is easy enough to grasp, sustained co-operation is difficult to achieve, even in the presence of widely acknowledged contract zones or zones of agreement. One of the most robust findings of the social sciences is reflected in the proposition that rational actors (in international society as well as elsewhere) engaging in interactive decision-making in the absence of effective rules or social conventions can and often do fail to realise feasible joint gains, thus ending up with outcomes that are suboptimal (and sometimes highly destructive) for all concerned.[2] While studies of the game-theoretic construct known as the prisoner's dilemma have come to symbolise these collective-action problems, analyses of a number of other prototypical cases of interactive decision-making yield the same conclusion.[3]

To this general proposition, we must add a number of considerations especially relevant to international society that complicate the problem facing those seeking to foster sustained co-operation at the international level. Since issues are seldom well-specified at the outset in international interactions, it is typically necessary to engage in one or more rounds of preliminary negotiations. This permits a number of things to be determined: the range of issues to be covered, the way in which the issues will be framed for purposes of negotiation and the identity of the actors who will participate in efforts to reach agreement on co-operative measures. The interactions in question normally include a number of actors who operate under a rule of unanimity in their efforts to reach agreement on the terms of co-operation, and thus there is always a danger that holdouts will seek better terms for themselves as the price for their participation in co-operative arrangements. The parties to international negotiations are complex collective entities, a fact that ensures the occurrence of extensive and sometimes crippling intra-party bargaining in the search for sustained international co-operation.[4] Further, the pursuit of co-operation in international society does not occur in a vacuum; one or more of the key participants will often be preoccupied with other issues (for example, elections or civil strife) to the point that they have difficulty focusing on the issues at hand.

The key to solving these collective-action problems, most observers now believe, lies in the creation of international regimes or, more broadly, international institutions.[5] Regimes are interlocking sets of rights and rules that govern interactions among their members with regard to particular issue areas. Given the diversity of international issues, it will come as no surprise that regimes vary greatly in terms of their membership, functional scope, geographical domain and administrative apparatus. Yet all regimes are social institutions, a fact that makes it possible to turn to the large body of literature on such institutions in endeavouring to understand international regimes and, particularly, international environmental regimes.[6]

Among students of international relations, great interest has developed over the last decade in the study of international economic regimes.[7] Many analysts have subjected the formation and evolution of the post-war trade and monetary regimes, for instance, to careful scrutiny. Now, there are indications that interest is growing rapidly in the study of international environmental regimes. At the leading edge of this rising tide of interest is the regime for the protection of stratospheric ozone set forth in the 1985 Vienna Convention for the Protection of the Ozone Layer, the 1987 Montreal Protocol on Substances that Deplete the Ozone Layer and the 1990 amendments to the Montreal Protocol. This focus is perfectly understandable, given the facts that the ozone layer regime is the first to deal with a highly visible problem of global environmental change, that the regime is notable as an anticipatory response to a future problem and that well-publicised efforts are already underway to extend and strengthen its provisions.[8] Yet other international environmental regimes are also becoming objects of increased attention among those concerned with global environmental changes, even though they are often more limited in their geographical reach.[9] Prominent among these are:

(1) The regime for air pollution established under the 1979 Geneva Convention on Long-Range Transboundary Air Pollution and its 1985 and 1988 protocols on sulphur emissions and nitrogen oxide emissions;

(2) The pollution control regime for the Mediterranean Basin set up under the 1976 Barcelona Convention for the Protection of the Mediterranean Sea Against Pollution and Its Related Protocols;

(3) The wildlife protection regimes set forth in the 1971 Ramsar Convention on Wetlands of International Importance Especially as Waterfowl Habitat, the 1973 Washington Convention on International Trade in Endangered Species of Wild Fauna and Flora and the 1979 Bonn Convention on the Conservation of Migratory Species of Wild Animals.[10]

As these observations suggest, many of those endeavouring to orchestrate international responses to global environmental changes have come to focus on a framework convention/protocol model of regime formation.

But this is not the only option available to those embarking on international co-operation as a means of coming to terms with large-scale environmental changes. The substantive provisions of many international environmental regimes are set forth in more or less complete form in initial agreements, so that there is no need to introduce a series of substantive protocols in quick succession following the adoption of a framework convention.[11] Cases in point include the 1946 International Convention for the Regulation of Whaling, the 1973/1978 MARPOL Convention for the Prevention of Pollution from Ships[12] and the 1980 Convention on the Conservation of Antarctic Marine Living Resources. Under the circumstances, a question regarding international environmental regimes which is worthy of additional study at this time concerns the circumstances under which one or the other of these models is most appropriate.

RESEARCH IN PROGRESS

So far, research on international environmental regimes and, for that matter, international institutions more generally has focused on processes of regime formation. The central concern of such studies is to identify the determinants of success or failure in efforts to form international regimes; some attention has been devoted as well to explaining the timing of successes in regime formation and to accounting for the content of the principal provisions of the regimes that form. Broadly, five streams of (not necessarily contradictory) analysis have emerged: structural arguments, power-based arguments, interest-based arguments, cognitive arguments and contextual arguments.

Structural analysis is based on the premise that not all forms of interactive decision-making are alike and that differences in the underlying character or structure of such relationships can facilitate or complicate efforts to reach agreement on the provisions of institutional arrangements.[13] The number of participants is important, it is argued, because the transaction costs of reaching agreement on the content of specific provisions rise steeply as the size of the group increases. The dimensions of the payoff or outcome space constitute another relevant variable. While virtually all cases of interactive decision-making involve a mix of incentives to compete and to co-operate, the nature of this mix in specific situations is an important determinant of success or failure in efforts to form international regimes. Similar observations are in order regarding the extent to which interactions are ongoing or iterative in character. Parties expecting to engage in a long running relationship are apt to find that the shadow of the future exerts considerable pressure on them to behave co-operatively at the outset. While the logic behind each of these propositions is straightforward, an initial assessment of the evidence regarding the formation of environmental regimes raises significant questions about their explanatory power.

The most prominent proposition of those who focus on power in analysing the formation of regimes is the argument, flowing from hegemonic stability theory, that the presence of a hegemon (in the sense of an actor possessing a preponderance of material resources) is necessary for regime formation.[14] With respect to international environmental regimes, at least, this proposition has not held up well in the face of empirical testing.[15] But this does not exhaust the potential linkages between power and regime formation; it is not difficult to see opportunities to formulate a number of other propositions relating structural power or the distribution of such power to regime formation and to apply them to the realm of environmental concerns.[16]

Those who think in terms of interests, by contrast, focus their attention on institutional bargaining[17] and on factors likely to impede or facilitate the negotiations involved in efforts to form regimes. From this perspective, the essential concern is to overcome the collective-action problems associated with all processes of interactive decision-making. Specific factors highlighted by this stream of analysis include: the extent to which negotiations lend themselves to integrative bargaining, the thickness of the veil of uncertainty,[18] the impact of exogenous shocks or crises and the role of leadership in overcoming the collective-action problems plaguing efforts to form international institutions.[19]

Cognitive explanations of regime formation take two principal forms. On the one hand, there is the argument that regimes are shaped by the power of ideas, a view that suggests that they can only come into existence in the presence of widely shared visions of the problems at stake and the appropriate solutions.[20] This way of thinking suggests the importance of hegemony in the Gramscian, in contrast to the material, sense in accounting for successes and failures in efforts to form international regimes.[21] On the other hand, another stream of analysis stressing cognitive factors highlights the emergence of an epistemic community (that is, a transnational group including officials and scientists who share both a set of causal beliefs and a set of normative preferences regarding the problems at hand) as a necessary condition for the achievement of success in efforts to form international environmental regimes.[22] While this argument requires further elaboration and may not stand up in its strongest form, there are indications that groups of this kind have played significant roles in the formation of some environmental regimes.

Contextual arguments provide a healthy reminder that regime formation in international society never takes place in a vacuum. In fact, efforts to devise institutional arrangements are closely tied to an array of political, economic and scientific concerns that can, and often do, accelerate or block processes of regime formation. For example, while the announcement of the ozone 'hole' over Antarctica in 1985 undoubtedly served to intensify public pressure to do something about the problem of ozone layer depletion,[23] the growing preoccupation of Soviet leaders with internal problems may make it increasingly difficult to negotiate international environmental regimes that include the Soviet Union as a participant. The underlying message here is that success in regime formation requires an ability to capitalise on windows of opportunity that emerge when a variety of contextual factors are aligned in ways that facilitate or even promote action with regard to the establishment of new institutions.

RESEARCH FRONTIERS

Given the fact that international environmental regimes are just beginning to come into focus as an important feature of international society, it is perhaps understandable that research on this type of response to global environmental changes has exhibited a preoccupation with regime formation. But there is now a compelling case for broadening the scope of these studies to include a variety of other concerns. Above all, it is apparent that we must improve our understanding of the effectiveness of institutional arrangements or, in other words, the extent to which the introduction of a regime affects or alters the behaviour of those subject to its provisions.[24] Unless environmental regimes can be shown to be effective in this sense, those concerned with mitigating or adjusting to global environmental changes will not continue for long to pay attention to the processes of regime formation or the requirements for operating regimes once they are in place.

Effectiveness is a function of implementation, compliance and institutional persistence. While research on this topic at the international level is in its infancy, some themes are already beginning to emerge. As students of implementation in other social settings have made clear, implementation is by no means a routine matter; institutional arrangements as implemented frequently diverge markedly from what a reading of the initial statute or agreement would lead one to expect.[25] Therefore, it is essential to look upon implementation as a political process and to analyse the hard bargaining that regularly takes place at this stage of the institution building process. Further, there is an important distinction to be drawn between compliance and enforcement.[26] Enforcement in the sense of the use of sanctions to influence the behaviour of subjects is one means of eliciting compliance. In most cases, however, the marginal costs of relying on enforcement to achieve compliance rise steeply. It follows that the effectiveness of institutional arrangements in all social settings depends heavily on the success of efforts to structure arrangements in such a way that those subject to a regime's rights and rules comply voluntarily. Institutional persistence is another complex matter. What stands out at this early stage is the extent to which inertia keeps institutional arrangements in place long after the constellation of interests giving rise to them has changed or even disappeared.[27] Viewed in this light, the lack of change in the institutional character of the United Nations Security Council, despite fundamental shifts in the distribution of power in international society, is no cause for surprise. Nor is the sudden collapse of institutional arrangements linking Eastern Europe and the Soviet Union unexpected; such occurrences are to be expected when the force of inertia is finally spent. As we move into a period of growing interest in the formation of international regimes as a response to global environmental changes, there can be no doubt that the importance of efforts to deepen our understanding of the determinants of institutional effectiveness will rise sharply.

There are, in addition, a number of new topics now coming into focus in connection with international environmental regimes; they will clearly merit more systematic investigation during the near future. For example, there is the matter of preparatory negotiations aimed at reaching agreement on appropriate ways to frame environmental problems at the international level and on the identity of the actors to be included as participants in the actual processes of regime formation. There is a world of difference, for instance, between endeavouring to devise a comprehensive law of the atmosphere (a conceptualisation which suggests parallels with the Law of the Sea) and seeking to establish separate regimes to deal with ozone layer depletion, global warming and acid rain. As a second example, there is also great dissimilarity between embracing the idea of creating a series of geographically limited or regional environmental protection regimes and opting for more comprehensive arrangements applicable to all regions, with suitable adjustments to deal with specific regional concerns. Among other things, the formulation of the problem at the preparatory stage goes a long way towards determining the selection of actors to participate in ensuing processes of regime formation. The selection of participants in institutional bargaining in turn plays a role of considerable importance as a determinant of success or failure in processes of regime formation.[28]

While those involved in devising international environmental regimes have so far managed to skirt the core issues dividing the advanced industrial countries and the countries of the developing world, the emerging agenda of global environmental changes is bound to bring us face to face with the basic sources of the North-South divide. Can large-scale environmental concerns be dealt with effectively in the absence of serious efforts to tackle the growth of human populations in the Third World? Why should developing countries, like China and India, agree to restrictions on the use of fossil fuels when the worst offenders in terms of greenhouse gas emissions are the advanced industrial states of the North? Do Northern states have any right to intervene in the affairs of developing countries, like Brazil and Indonesia, whose actions are critical to accelerating or stemming the loss of biodiversity? What is required in this context is a sustained effort to understand the politics of regime formation and regime effectiveness when the institutions in question must encompass the concerns of a deeply divided set of participants. The compensation fund to be set up under the 1990 amendment to the Montreal Protocol is particularly interesting in this context. It reminds us, for one thing, of the need to think carefully about the bargaining power of apparently weak players, like China, which can issue credible threats to step up their consumption of CFCs or fossil fuels unless others make it worthwhile for them to desist, and Brazil, which can do the same with respect to the destruction of moist tropical forests.

The emerging role of non-state actors – including intergovernmental organisations, nongovernmental organisations and transnational corporations – in the formation and operation of international regimes is another subject meriting more systematic investigation. The environmental issue area lends itself to active participation on the part of non-state actors. This does not mean that states will soon be displaced or even supplemented by non-state actors as the formal participants in processes of regime formation or in the operation of institutional arrangements. Rather, non-state actors are increasingly consulted by those negotiating the terms of environmental regimes on behalf of states, represented on national delegations to such negotiations and allowed to participate informally but effectively in the operation of institutions. What is emerging, at least in this issue area, is a more complex international society in which states remain important actors, but find themselves increasingly sharing influence, if not authority, with several other types of actors. To the extent that this trend continues and spreads, the need for more sophisticated conceptualisations of international interactions will become increasingly compelling.

There is much to be said also for an expanded effort to think through the relationships between institutions and organisations in efforts to achieve and sustain the international co-operation needed to deal with global environmental changes. Institutions are interlocking sets of rights and rules that serve to guide the interactions of parties in specific issue areas. Organisations, by contrast, are material entities possessing offices, personnel, equipment, budgets and legal personality.[29] The regimes for the protection of the stratospheric ozone layer and the control of pollution in the Mediterranean Basin are institutions; the United Nations Environment Programme (UNEP) and the World Meteorological Organization are organisations. In some areas, organisations have become important players in processes of regime formation at the international level; there is a strong case to be made, for instance, for the proposition that UNEP assumed such a role in connection with the formation of both the pollution control regime for the Mediterranean and the ozone layer regime. On the other hand, some (but not all) institutions require the services of organisations to manage or administer them. Though implementation of the key rules regarding production and consumption of CFCs is delegated to the governments of the members states in the case of the ozone layer regime, for example, the newly established compensation fund will necessitate the creation of a new organisation. The international whaling regime, to go a step further, requires administrative mechanisms to compile statistics regarding whales and whaling, make determinations regarding the status of whale stocks, arrive at decisions about annual catch quotas (if any) and monitor compliance with its decisions. Given the growing role of international environmental regimes, coupled with the fact that organisations are costly to operate in both monetary and non-monetary terms, the value of a better understanding of the relationships between institutions and organisations is evident for those charged with coming to terms with global environmental changes.

CONCLUSION

There is no escaping the need for sustained international co-operation as one component of the overall human response to global environmental changes. Because collective-action problems are just as pervasive in this issue area as they are elsewhere, international co-operation relating to environmental matters requires the formation and operation of an array of regimes. These regimes in turn constitute a subset of the broader class of social institutions. Accordingly, the study of international environmental regimes can not only profit from the application of propositions relating to social institutions more generally, but it can also contribute to the growth of knowledge about social institutions. While much remains to be learned about the processes of regime formation in international society, there is now a compelling case for broadening the scope of the enquiry to focus on factors determining the effectiveness of international institutions, including environmental regimes. Additionally, the study of such arrangements offers opportunities to shed light on a number of related topics, including preparatory activities leading up to institutional bargaining, the problems associated with North-South interactions, the role of non-state actors and the relationships between institutions and organisations.

Notes

This chapter draws heavily upon work in progress under the auspices of the Committee on the Human Dimensions of Global Change of the United States National Academy of Sciences.

1. Malcolm W. Brown, '93 Nations Agree to Ban Chemicals that Harm Ozone', *The New York Times*, 30 June 1990, pp. 1–2.
2. Among those interested in environmental matters, this phenomenon is often discussed in terms of the 'tragedy of the commons' metaphor. See Garrett Hardin, 'The Tragedy of the Commons', *Science* (Vol. 162, 1968) pp. 1243–8.
3. Russell Hardin, *Collective Action* (Baltimore, MD: Johns Hopkins University Press, 1982).
4. Robert D. Putnam, 'Diplomacy and Domestic Politics: The Logic of Two-level Games', *International Organization* (Vol. 42, Summer 1988) pp. 427–60.
5. Stephen D. Krasner (ed.), *International Regimes* (Ithaca, NY: Cornell University Press, 1983); Oran R. Young, *International Cooperation: Building Regimes for Natural Resources and the Environment* (Ithaca, NY: Cornell University Press, 1989); and Robert O. Keohane, *International Institutions and State Power* (Boulder, CO: Westview, 1989).
6. Beth V. Yarborough and Robert M. Yarborough, 'International Institutions and the New Economics of Organizations', *International Organization* (Vol. 44, Spring 1990) pp. 235–59.
7. Richard N. Cooper, *The International Monetary System: Essays in World Economics* (Cambridge, MA: MIT Press, 1987); Charles P. Kindleberger, *The International Economic Order: Essays on Financial Crisis and International Public Goods* (Cambridge, MA: MIT Press, 1988); and Robert O. Keohane, *After Hegemony: Cooperation and Discord in the World Political Economy* (Princeton, NJ: Princeton University Press, 1984).
8. Richard Elliot Benedick, *Ozone Diplomacy* (Cambridge, MA: Harvard University Press, 1991).

9. Peter H. Sand, 'International Cooperation: The Environmental Experience', in Jessica Tuchman Mathews (ed.), *Preserving the Global Environment: The Challenge of Shared Leadership* (New York: W.W. Norton, 1990).
10. Simon Lyster, *International Wildlife Law* (Cambridge: Grotius Publications, 1985).
11. Peter Thacher, who participated on behalf of the United Nations Environment Programme during the Mediterranean Action Plan discussions, has observed that the strategy of those who negotiated the pollution control regime for the Mediterranean was to devise a framework convention and several protocols simultaneously and to require states signing the framework convention to adhere to at least one of the protocols at the same time.
12. The MARPOL Convention does, however, contain a number of annexes that can be ratified or rejected on a case by case basis.
13. Kenneth A. Oye, 'Explaining Cooperation Under Anarchy: Hypotheses and Strategies', in Kenneth A. Oye (ed.), *Cooperation Under Anarchy* (Princeton, NJ: Princeton University Press, 1986) pp. 1–24.
14. Keohane, *op. cit.*, in note 7, Chapter 3.
15. For an argument that the proposition is also suspect on theoretical grounds, see Duncan Snidal, 'The Limits of Hegemonic Stability Theory', *International Organization* (Vol. 39, Autumn 1985) pp. 579–614.
16. On the idea of structural power, see Stephen D. Krasner, *Structural Conflict: The Third World Against Global Liberalism* (Berkeley, CA: University of California Press, 1985); and Susan Strange, *States and Markets: An Introduction to International Political Economy* (New York: Basil Blackwell, 1988).
17. Institutional bargaining refers to efforts on the part of autonomous actors to reach agreement on the terms of constitutional contracts or interlocking sets of rights and rules that are expected to guide their subsequent interactions in a given issue area.
18. On the concept of a 'veil of uncertainty' and the differences between this idea and Rawls' 'veil of ignorance', see Geoffrey Brennan and James M. Buchanan, *The Reason of Rules: Constitutional Political Economy* (Cambridge: Cambridge University Press, 1985) Chapter 2.
19. Oran R. Young, 'The Politics of International Regime Formation: Managing Natural Resources and the Environment', *International Organization* (Vol. 43, Summer 1989) pp. 349–75.
20. For an account of the power of ideas in international society, see Ernst B. Haas, *When Knowledge is Power: Three Models of Change in International Organizations* (Berkeley, CA: University of California Press, 1990).
21. Robert W. Cox, 'Gramsci, Hegemony and International Relations: An Essay in Method', *Millennium: Journal of International Studies* (Vol. 12, No. 2, Summer 1983) pp. 162–75.

22. See Peter M. Haas, 'Obtaining International Environmental Protection through Epistemic Consensus', in Chapter 4.
23. Sharon L. Roan, *Ozone Crisis: The 15–Year Evolution of a Sudden Global Emergency* (New York: John Wiley, 1989).
24. Oran R. Young, 'The Effectiveness of International Institutions: Hard Cases and Critical Variables', in James N. Rosenau (ed.), *Governance without Government in International Society* (Cambridge: Cambridge University Press, forthcoming).
25. Jeffrey L. Pressman and Aaron Wildavsky, *Implementation: How Great Expectations in Washington Are Dashed in Oakland*, Second Edition (Berkeley, CA: University of California Press, 1973).
26. Oran R. Young, *Compliance and Public Authority: A Theory with International Implications* (Baltimore, MD: Johns Hopkins University Press, 1979); and Roger Fisher, *Improving Compliance with International Law* (Charlottesville, VA: University Press of Virginia, 1981).
27. Stephen D. Krasner, 'Sovereignty: An Institutional Perspective', in James A. Caporaso (ed.), *The Elusive State: International and Comparative Perspectives* (Newbury Park, CA: Sage, 1989) pp. 69–96.
28. These considerations account for the high level of interest currently being expressed in the work of the Preparatory Committee for the 1992 United Nations Conference on Environment and Development.
29. Young, *op. cit.*, in note 5, Chapter 2.

3 The International Politics of Global Environmental Change

Ian H. Rowlands

INTRODUCTION

The issue of global environmental change entered the international agenda during the mid and late 1980s. Crises events – whether manifestations of humankind's activities or natural occurrences – served to illuminate the stresses that human activities were placing upon the earth's natural systems. High-profile media coverage and mounting public pressure helped to prompt national leaders to turn their attentions towards the global problems of ozone layer depletion and climatic change. Working parties were established, commissions were formed, conferences were convened and reports were tabled. The world experienced an unprecedented flurry of international activity on environmental issues.

The purpose of this chapter is to provide an overview of the international dimensions of ozone layer depletion and global warming. To achieve this end, I will look at the two issues in turn and undertake four tasks in each section. First, I will provide a basic description of the natural processes involved. Second, I will recount what are purported to be the most significant impacts of these natural changes upon social systems. Third, I will review the history of international political activity on the issues. And finally, I will offer some preliminary thoughts about the major issues that may lie ahead for international society.

THE OZONE LAYER DEPLETION ISSUE

The Natural Processes and the Possible Impacts

The ozone layer, located 10 to 50 kilometres above the earth's surface, absorbs most of the incoming ultraviolet radiation. The natural equilibria of the gases in this layer have been disrupted recently by a number of human-made substances that have been destroying the ozone. If these chemicals (the major ones being chlorofluorocarbons [CFCs]) continue to be emitted and the world experiences a depletion of its protective ozone layer, then more ultraviolet radiation will be able to penetrate the atmosphere and hit the earth's surface. This increase in ultraviolet radiation would have devastating effects upon humans, animals, material objects and the natural environment.

The US Environmental Protection Agency (EPA) in 1987 completed a study that examined the implications of increased ultraviolet radiation (that is, if nothing were done to save the ozone layer) for the US population. The report found that among those people alive today and born by 2075, there would be an additional 163 million to 308 million cases of skin cancer, of which 3.5 million to 6.5 million of these cases would be fatal.[1] Under the same scenario and taken worldwide, it has been suggested that 'skin cancer increases alone could exceed *1–2 billion* cases over this period'.[2] It has also been shown that there would be an increase in the incidence of cataracts and a general weakening of the immune system, making all people more susceptible to illness and disease. Further, similar health ailments would affect animals. With greater amounts of ultraviolet radiation, therefore, the productivity of raising farm animals would decrease. Perhaps more significantly, preliminary studies suggest that the DNA of phytoplankton – the tiny sea-organisms which photosynthesise light and form the basis of the food chain – would be damaged. Declining stocks of phytoplankton would place the existence of their predators in doubt and, like dominoes, endanger the whole food chain. Additionally, research has indicated that some plants would be harmed by greater doses of ultraviolet radiation, and thus crop yields would decline. Non-living things would be also affected by a thinner ozone layer. A variety of manufacturing materials would be weakened by greater ultraviolet flux and therefore would have to be replaced more frequently. Finally, because CFCs are also 'greenhouse gases', they would not only deplete the ozone layer, but they would also accelerate the rate of climatic change.[3]

International Political Activity

Although ozone was first detected late in the nineteenth century, the presence of ozone in the stratosphere was not discovered until 1917. Speculation about the formation and destruction of ozone prompted British chemist Sydney Chapman to investigate the substance. In 1930, he proposed that the amount of ozone present was dictated only by the concentrations of atomic oxygen, molecular oxygen and ozone. His work became the accepted foundation of this branch of atmospheric chemistry for many years.

Following an investigation of new data collected after the International Geophysical Year (1957–58), however, scientists learned that other atmospheric trace gases also affect the ozone balance. The discovery that both hydrogen and nitrogen compounds influence ozone levels helped to launch the public debate about the ozone layer.[4]

This new scientific information was used by environmentalists (and others) in order to challenge the supersonic transport (SST) project, which was being developed during the late 1960s.[5] They claimed that the substances released by the airplanes while in flight would threaten the ozone layer. One estimate suggested that there could be 500 SSTs flying by 1985 (each of them flying on average seven hours a day), leading them to argue that the damage could be significant. Although at the time most of the development of the SST was being undertaken by Boeing in the US, the Soviets were also building a prototype (the Tupolev 144) and the French and the British were collaborating on the Concorde. In light of the scientific allegations, the US government decided to investigate further the possible dangers of SST flights, and the US Congress ordered a report in 1971. This report, initiated by the US Department of Transport and undertaken by the Climate Impact Assessment Program, was completed in 1974 at a total cost of US$40 million.[6] The final report concluded that a 500-plane fleet of Boeing SSTs would have caused a 16 per cent depletion of ozone in the northern hemisphere and an 8 per cent depletion in the southern hemisphere.[7] But by this time, Boeing's programme had been discontinued. (The Senate, in 1971, had cancelled the plan. Not only was the environmental impact of its operation being questioned, but the SST's economic non-viability also played a major role in the final decision.) The Soviet programme was also eventually cancelled. Further, the commercial viability of the Concorde never materialised, and today there are only about a dozen Concordes in operation. The SST controversy, however, marked the first time that the issue of stratospheric ozone had come upon the agenda of any national government. And indeed the ensuing debate about US landing rights for the Concorde demonstrates that this issue also had international dimensions.

During the early 1970s, there were two other anthropogenic threats to the ozone layer that were recognised as being potentially significant. The first was the effects of nuclear explosions. This issue would have received greater attention if international agreements to restrict above-ground nuclear tests had not been concluded in the 1960s. At this time, it was noted that the 'effects of nuclear bomb tests around 1960 on ozone [were] only marginally detectable'.[8]

The other acknowledged threat to the ozone layer related to agricultural practices. The extent to which nitrogen compounds (which are released from the use of fertilisers) could deplete stratospheric ozone was starting to be considered. The potential impact was still, however, a matter of much debate.

These two threats did not have a major international impact, and with the waning of the SST issue, public interest in the ozone layer subsided. However, the issue once again started to occupy a part of the public's consciousness after the 1974 publication of a paper in the scientific journal *Nature*.[9] In this paper, it was hypothesised that the chlorine present in CFCs had the potential to destroy significant amounts of stratospheric ozone. With this supposition, the international ozone layer debate entered a new dimension.

The scientific community was in an uproar as controversy over the validity of the theory ensued. But because it was such a serious assertion, with significant implications, it was not taken lightly. A major international conference was convened in Washington, DC in March 1977. Its delegates produced a World Plan of Action on ozone that called for greater monitoring and research into both technical and social issues in the problem area.[10]

The worldwide political response was varied, and by the late 1970s, two blocs had formed. On the one hand, the United States (soon to be joined by Sweden, Norway and Canada) had outlawed the use of CFCs in non-essential aerosols in 1978 and was calling for a global ban. On the other hand, the major states of the European Economic Community, along with Japan, had refused to impose stringent controls. Citing uncertainties in the theory and the lack of empirical evidence, the European Economic Community only called upon member states to reduce non-essential aerosol usage voluntarily by 30 per cent of their 1976 production figure by 1982.[11]

Neither side wavered from its position, and during the early 1980s little political movement took place. Further, governments' interest in the issue diminished during the early 1980s for four major reasons. First, in 1981 the new Reagan Administration in the US appointed Anne Gorsuch to head the EPA. She did not pursue the issue vigorously, because she dismissed ozone depletion as just another environmental scare.[12] Second, the international scientific group that was studying the issue was steadily revising its calculation of ozone depletion – *downward*. It estimated eventual ozone depletion to be 15 per cent in 1979, but only 10 per cent in 1980, and down to between 5 and 10 per cent in 1981.[13] Third, the world was experiencing a recession, and environmental matters took a backseat to economic issues. Finally, owing mainly to the US 'can ban', the use of CFCs was declining worldwide.[14]

Nevertheless, under the auspices of the United Nations Environment Programme (UNEP), international negotiations towards a convention and protocol commenced in January 1982.[15] Just over a year into the process, international interest in the issue was revived: there was a change in the EPA leadership, there was increased growth in the use of CFCs and the chemicals' potential link to global climatic change was firmly established. The friction between the two groupings of countries, however, endured.

Although negotiations for a convention were proceeding satisfactorily, conflict inhibited the prospects for an accompanying protocol. The Americans, the Scandinavians and the Canadians demanded that an aerosol ban be written into the ozone convention. The Europeans, for their part, did not want to focus solely on aerosols, because they believed that this approach would be beneficial only in the short term. They proposed a production cap on all CFCs. The dispute persisted.[16]

On 22 March 1985, the 'Vienna Convention for the Protection of the Ozone Layer' was signed by 20 countries.[17] There were no specific obligations upon the parties to the convention. Rather, it created a framework for international co-operation on research, monitoring and information exchange with respect to the ozone layer, potential modification of the ozone layer and the potentially adverse health, environmental and climatic effects of such modification. Nevertheless, this was a very historic document, because, for the first time, states agreed in principle to tackle a global environmental problem before its effects were felt or its scientific foundations firmly proved. With the convention signed, the next step was to try to achieve a protocol.

In the middle of 1985, the international efforts were given a further sense of urgency by two developments. First, the discovery of a significant 'crater' in the ozone layer above the Antarctic was reported by the British Antarctic Survey.[18] Second, a report was jointly released by the US National Aeronautics and Space Administration (NASA) and UNEP in July of 1985. In this report, 150 scientists from 11 countries concluded that the ozone layer had already been damaged. However, they also stated that they had too little information to predict what the future might hold.[19]

By the summer of 1986, a consensus had been reached upon a number of issues: that substitutes were limited by price, not chemistry; that production rates were rising; that the concentration of the chemicals in the atmosphere was increasing; and that emissions must be cut by 85 per cent to keep chlorine levels from growing.[20] Some progress was made at a workshop in Leesburg, Virginia, US in September 1986. A change of heart was evident, because, working upon a Canadian-proposed compromise, a broad agreement in favour of a cap on global emissions, rather than on the restriction of end uses, emerged.[21] UNEP itself has noted that

> the Leesburg meeting considered various options for controls, but took care not to contrast between them. In the friendly atmosphere important concessions were made, trust was built up, and for the first time an obvious international will to forge a successful protocol emerged.[22]

With the last informal technical workshop concluded, the formal negotiations on a control protocol began in Geneva in December 1986. Negotiations towards a protocol continued fervidly during 1987, and acceptable proposals were slowly being agreed upon. Finally, on 16 September 1987, 27 countries signed the 'Montreal Protocol on Substances that Deplete the Ozone Layer'.[23] The significance of this document is that it committed signatories to reduce their consumption of certain CFCs by 50 per cent of their 1986 figure by 1999. The wording of the final document reflects the delicacy of the negotiations, because it contains a number of clauses to cover the special circumstances of several states. For example, the Europeans insisted that the limits be placed on consumption *and* production, to protect their export markets. Further, concessions were made to allow existing industrial producers in the US, the European Community (EC) and Japan to produce up to 10 per cent more if the incremental production went to developing countries. (This was to discourage developing countries from constructing their own CFC-production facilities.) The Soviet Union was also allowed to complete two CFC plants then under construction and to increase per capita consumption (to 0.5 kilogrammes/capita) so as to account for the implementation of its five-year plans. Finally, states of the developing world were given a ten-year period of grace to implement the controls.[24] These facts should not, however, diminish the significance of the document.

Just as the ink was drying on the Montreal Protocol, an important scientific expedition set off from Punta Arenas in Chile. With 150 scientists and support staff from 19 organisations, this NASA-sponsored expedition travelled to the Antarctic in order to investigate ozone depletion. The group's discoveries demonstrated 'an undoubted chemical cause in the destruction of ozone by atmospheric chlorine'.[25] These results, which implicated CFCs, not only highlighted the significance of the just-signed protocol, but also suggested that its controls were perhaps not strong enough.

During the entire 14-year history of the CFC-debate, the chemical industry had been adamant in its belief that there was not enough scientific evidence to warrant international controls. Thus Du Pont's announcement, in March 1988, that it planned to phase out CFCs was quite notable.[26] It is generally agreed that this decision was motivated by the results of a NASA-sponsored study, released in the same month. Following upon the autumn expedition to the Antarctic, the NASA-sponsored group (which had over 100 scientists from 10 countries) had scrutinised a large amount of data and found ozone depletion in excess of what had been predicted by computer models.[27]

In 1988, such mounting evidence of ozone depletion prompted a number of national governments, including Sweden, Norway, Finland and the Netherlands, to take unilateral control action beyond their obligations as outlined in the Montreal Protocol. Further, in March 1989, the environment ministers of the EC agreed to phase out CFC use by 85 per cent as soon as possible and to seek a total ban by the end of the century. The next day, US President George Bush stated that the US would join the ban. Thus, in a space of two days, 13 countries, among them producing over two-thirds of the world's CFCs, had agreed to a total phase-out of the chemicals. An unprecedented agreement about the severity of the problem was emerging among the industrialised states. This consensus was one of two dominant themes that became prominent at major international conferences in London in March 1989 and in Helsinki in May 1989.

At these meetings, a general sense of urgency prevailed as delegates recognised that the provisions contained in the Montreal Protocol would not adequately address the ozone layer problem. A feeling was emerging that the Protocol would have to be amended and that the timetable for reducing and eventually eliminating CFCs would have to be brought forward.

But delegates were also accepting the fact that the Montreal Protocol would have to be altered in another manner. At these two meetings, the issue of global equity became dominant. Politicians from developing states, led by the Chinese and Indian representatives, demanded assistance in order to meet the obligations of the Montreal Protocol.[28] They argued that because the industrialised world had caused most of the destruction of stratospheric ozone, the developed countries should be primarily responsible for paying the costs of repair. The developing states' leaders made it clear that they did not want to jeopardise their prospects for a higher level of development by foregoing the use of these chemicals. They called for technology and financial resources to be transferred – free of charge – from the North to the South. This would allow their citizens to leap-frog the use of CFCs and immediately use substitute chemicals.

Decision-makers from the North recognised that Southern participation in the Protocol was essential to its success. Although the developing world produces a relatively small amount of CFCs, it was accepted that it could easily expand its capacity for production.[29] Thus these demands became a primary issue in the international politics of the ozone layer.

In the middle of 1989, a number of working groups were formed in order to investigate proposals to amend the Protocol, with the questions of stricter controls and an international fund being at the top of their agendas. Negotiations were, however, soon frustrated by a conflict over the funding mechanism. The Europeans moved to support its creation, but the US opposed it, worrying about the precedent that such a fund might set for the larger issue of global warming. Meetings in May 1990 in Geneva and Bergen were hampered by this conflict.[30]

The US, however, shifted its position in June 1990 and agreed to the principle of an international fund. Nevertheless, some issues were still left unresolved (for example, the fund's size and administration) as the parties to the Montreal Protocol met in London in June 1990 in order to amend the agreement legally. After three days of intense negotiations at the ministerial level, some consensus was finally reached. First, the delegates agreed to phase out CFCs completely by 2000. Further, controls on halons were strengthened and controls on carbon tetrachloride and methyl chloroform were introduced. Additionally, a US$240 million fund was established.[31] This money, to cover an initial three-year period, would be used to assist developing countries to switch from CFCs. In this way, less than three years after the Montreal Protocol had been written, the terms of the ground-breaking document were considerably tightened.

Towards the Future

Although there has been considerable success in the international efforts to protect the ozone layer, we should not assume that the case is by any means closed. As of the end of 1990, India and China, despite signs of interest, had yet to join the procedure. Thus, as was noted before, any chance for a truly global effort to combat the issue would be defeated if these two huge countries, with their great potential to produce ozone-destroying chemicals, were to remain outside the Protocol.

Further, a number of scientists and pressure groups argue that the enhanced regulations agreed in London do not go far enough. Even with the new controls, scientists say that the rate of ozone layer depletion will continue to rise until at least 2000, and that the Antarctic ozone crater will not be fully repaired until well into the second half of the twenty-first century. Thus it is apparent that they believe that the deleterious effects of ozone layer depletion could still threaten international society. As a result, numerous scientists and policy analysts have called for even stricter controls and an accelerated timetable.

Third, the substances that are replacing the CFCs are quickly becoming topics of debate. It has been argued that the hydrochlorofluorocarbons (HCFCs, the most significant substitute chemicals), although more benign than CFCs, will nevertheless destroy some stratospheric ozone. Because chemical companies estimate that these 'second generation' ozone depletors will replace about 30 per cent of the CFC market, pressures are building for restrictions to be imposed. In London, however, no legal controls were placed upon HCFCs, although a declaration that they should be used carefully by industry and phased out by between 2020 and 2040 was agreed.[32] With such a long time horizon, the international political dimensions of the issue may well persist.

Finally, just as Thomas Midgely had no idea in 1930 that the new cooling agent which he had just invented would deplete stratospheric ozone, there may be other substances, presently being used, that could do the same. At the June 1990 meeting in London, delegates agreed to review the Protocol again in 1992, 'to see if the timetable could be improved'.[33] If the international community is to address adequately the challenges of the ozone layer depletion issue, its members must remain vigilant.

THE GLOBAL WARMING ISSUE

The Natural Processes and the Possible Impacts

The earth's atmosphere is constituted so that it allows most sunlight to stream in uninterrupted. After striking the earth's surface, this solar energy is reflected as longer-wavelength infrared radiation. Some of this radiation is subsequently trapped in the atmosphere by clouds and 'greenhouse gases' (which include carbon dioxide, methane, nitrous oxide and CFCs). Without this greenhouse effect, the surface of the earth would be about 33°C cooler than it presently is, and thus the phenomenon is necessary for life, as we know it, on the planet. But since the Industrial Revolution, humankind has pumped more gases into the atmosphere and has, in effect, 'thickened' the greenhouse blanket that surrounds the earth, therefore trapping more heat near the surface. The Intergovernmental Panel on Climate Change (IPCC) has estimated that average global temperature may rise by 1°C by 2025 and by 3°C by the end of the twenty-first century. This increase in average global temperature would induce a number of natural changes that would have significant consequences for the world's population.

One major outcome of higher temperatures would be a rise in sea levels. Given that warmer water occupies more volume, it has been estimated that the world's oceans could rise by between 10 cm and 30 cm by 2030 and by between 30 cm and 100 cm by 2100. Flooding in coastal areas would cause a recession of shorelands and wetlands, displace low-lying urban infrastructure and increase the intrusion of salt into fresh-water supplies. A shortage of freshwater – for both agricultural activities and human consumption – would result.

In a warmer world, the resulting shift in climatic zones would also have a significant impact. First, with different climatic characteristics for a given geographical location, agricultural patterns would be forced to change. With human intervention, some degree of adaptation may be feasible on private lands, but changes in the 'wilds' would have significant ecological consequences. Researchers suggest that not only would certain species vanish, but entire specific ecotypes could also be lost. Further, forest areas would shrink, because trees would not be able to migrate quickly enough to keep up with the shifting climatic zones. Those living things that would prosper in a warmer world would be weeds and insect pests – thus causing more havoc for human settlements and other living creatures.

Additionally, all regions of the world would experience a harsher and more unpredictable climate with greater incidence of storms, floods and droughts. Considering all of these developments, the 'IPCC Working Group 2 report emphasises that the regions that appear to be at greatest risk for "sustaining the population" are those that are already arid and marginal',[34] suggesting that the world's poor would be hardest hit. There is little doubt that climatic change would give rise to a greater number of environmental refugees – people driven off their land by direct or indirect environmental change.

International Political Activity

The greenhouse effect was first described by the French mathematician Baron Jean Baptiste Fourier in 1827. In 1896, Svante Arrhenius, a Swedish scientist, published a paper which postulated that an effective doubling of the amount of carbon dioxide in the atmosphere would cause the average global temperature to rise by 5°C.[35] Subsequent work in the area has effectively borne out his conclusions.

Yet the question of global warming did not reach international headlines until the 1980s. Instead, during much of the post-World War II period, the international community – if interested in climatic change at all – was concerned with global cooling. Average global temperatures had decreased from 1945 to 1970, and this had initiated a spell of investigations into the prospect of a coming ice-age.

One of the first serious inquiries into the possibility of global warming was held at the Massachusetts Institute of Technology (MIT) in July 1970. Researchers attending a conference on 'The Study of Critical Environmental Problems' concluded that the likelihood of climatic change during the twentieth century was small. However, they did 'not discount the possibility of such consequences in the longer term, and [they] recommended continuous measurement of the CO_2 content of the atmosphere'.[36]

The World Meteorological Organization (WMO) convened the First World Climate Conference in Geneva in February 1979. Much of the discussion centred around the debate between the ice-age prophets and those who saw global warming as being more important in the medium-term. It seemed that a consensus was slowly building in favour of the latter view. The final declaration agreed that

> we can say with some confidence that the burning of fossil fuels, deforestation, and changes of land use have increased the amount of carbon dioxide in the atmosphere . . . and it appears plausible that [this] can contribute to a gradual warming of the lower atmosphere, especially at high latitudes . . . It is possible that some effects on a regional and global scale may . . . become significant before the middle of the next century.[37]

Thus, just over a decade ago, it seemed that global warming would not become a political issue before the twenty-first century.

In October 1985, scientists from 29 countries met in Villach, Austria in order to review the then-current state of scientific understanding about the greenhouse effect. Two further workshops were held in Villach and Bellagio, Italy in 1987. 'The Villach-Bellagio workshops were significant for two main reasons. Not only were global warming trends confirmed, but, for the first time, policies to respond to climatic change were discussed in detail and agreed.'[38] The question of global warming subsequently entered the international agenda in 1988. In June 1988, the Toronto Conference on 'The Changing Atmosphere' was held. With over 300 participants from 46 countries, it was the first major international gathering to focus on global warming. The final conference declaration stressed the need for a comprehensive global convention as a framework for protocols on the protection of the atmosphere. The delegates also proposed a World Atmosphere Fund as a financial mechanism to assist the developing states. More concretely, they also called for a 20 per cent reduction from 1988 levels of carbon dioxide emissions by 2005.[39]

In North America, the summer of 1988 will also be remembered as one of the hottest on record. The drought conditions, although not necessarily manifestations of global warming, galvanised interest in the issue for both policy-makers and the general public. Additionally, an influential statement was delivered on 23 June 1988: James Hansen of NASA's Goddard Institute for Space Studies in New York appeared before the US Senate Energy Committee and declared that he was 99 per cent certain that the warming of the 1980s was not a chance event. He went on to argue that it was time to stop stalling on the basis of scientific uncertainty and time to start taking action to address global warming. This statement attracted widespread attention and illuminated the importance of the issue.[40]

On 6 December 1988, the UN General Assembly, at the initiative of the Government of Malta, passed a resolution (43/53) that formally requested the UNEP and the WMO, through the IPCC, 'immediately to initiate action leading, as soon as possible to a comprehensive review and recommendations with respect to . . . elements for inclusion in a possible future convention on climate'.[41] The assembly also recognised that climate change was the Common Concern of Mankind. Thus, with two international organisations committed to the issue, climatic change became the subject of global debate.

During 1989 and 1990 there were numerous international conferences on the issue. Let me just refer to a couple of the most significant meetings. In March 1989, the governments of the Netherlands, France and Norway hosted an international conference in The Hague. Representatives from 24 countries attended and issued a declaration that emphasised the use of legal instruments in trying to stabilise the atmosphere. The participants also envisaged an important role for the International Court of Justice, namely that of dispute resolution.[42]

At the World Economic Summit (G7) meeting in Paris that July, the leaders of the world's largest industrialised states addressed 'green' issues for the first time as a group. Their final communique recognised that 'decisive action is urgently needed to understand and protect the earth's ecological balance'.[43]

Meanwhile, at the other end of the economic spectrum, Rajiv Gandhi (the Indian Prime Minister at the time) called for a huge transfer of resources from the North to the South. Making the proposition at a meeting of the Non-Aligned Movement in Belgrade in September 1989,

Gandhi outlined his plan for a Planet Protection Fund, putting the price at US$18 billion.[44]

A major ministerial conference on atmospheric pollution and climate change was held in Noordwijk, the Netherlands, in November 1989. Although environment ministers from 68 countries were unanimous in their call for a climate convention as soon as possible, more specific agreement could not be reached. The United States, supported by the Japan and the Soviet Union, proved to be the most resistant to any further declarations.[45] They refused to agree to a Dutch proposal that would limit current levels of carbon dioxide emissions by the turn of the century. Thus, the battle-lines in this international debate were, at this time, firmly established.

This confrontation continued at a conference hosted by the United States in April 1990. Following George Bush's election promise to implement the 'White House Effect', the President gathered international decision-makers and experts on the global warming issue. But once again, emphasis was placed upon further research and upon, in the meantime, a 'business-as-usual' approach. The meeting ended in disappointment, because delegates resented the Bush Administration's attempts to recruit their governments to this passive position.[46]

The United States continued to occupy the role of chief antagonist at a meeting in Sundsvall, Sweden in August 1990. At this conference, representatives from 75 countries drew up the final report of the IPCC. Throughout the proceedings, US representatives frequently tried to tone down statements by citing scientific uncertainty.[47] This notwithstanding, the delegates finalised the report.

The IPCC's final report was then presented at the Second World Climate Conference in Geneva in November 1990. At this conference, delegates from 137 countries agreed that 'within 20 months there should be a new international law aimed at preventing potentially catastrophic man-made changes in climate'.[48] Formal negotiations began in Washington, DC in February 1991, with subsequent meetings scheduled to be held in Geneva and Nairobi. In this way, the international process is entering its next dimension. The aim is to have a convention ready for signing by 1992.

Towards the Future

It appears that the global warming issue will remain on the international political agenda during the early 1990s. Although many believe that unprecedented actions will have to be undertaken in order to address adequately the issue, it is still unclear how international society will deal with the problems. From a wide range of concerns, four issues seem to pose the largest challenges.

First, scientific uncertainty may continue to play a role in the international politics of global warming. The underlying theory of the 'greenhouse effect' is generally accepted. Further, 'the IPCC assessment conclude[d] first that "we are certain that increased emissions of greenhouse gases will result in additional warming of the earth's surface".'[49] There is, however, no consensus about the magnitude of the effect and the severity of the impacts. Additionally, the associated distributions – because not every area will be affected equally – are still being debated. In this way, scientists today are dealing 'with risks and probabilities, which are no less real or dangerous for being uncertain'.[50] Without '100 per cent' statements, however, policy-makers may be able to continue to stall the negotiations (as the Americans and occasionally others have in the past) on the basis of scientific uncertainty.

Second, one of the major hurdles impeding the implementation of any global convention on climate change is the perception that any action will be expensive. Figures in the billions and trillions of US dollars are often mentioned when the costs of adjusting energy consumption are calculated. Although there are those who challenge this assertion – some argue that it will actually result in a net economic benefit – the popular perception in many states is that both restructuring domestic society and assisting those in the developing world will cost dearly. Further, with a global economic recession perhaps just around the corner, the priority of environmental issues may be downgraded.

Third, just as North-South questions emerged as important issues in the international politics of the ozone layer, there is no reason to doubt that they will do the same in the global warming issue. At the Second World Climate Conference in Geneva in November 1990, UNEP Executive-Director Mostafa Tolba called on industrialised states to help poorer ones.[51] The final declaration also recognised that 'third world countries are bound to increase their greenhouse gas emissions as they industrialise'.[52] Further, 'it recommend[ed] they be given extra financial aid and technology to help minimise the environmental impact of industrialisation'.[53] For the same reasons as cited earlier, this issue of 'fairness' will have to be addressed – to the satisfaction of all major participants – before an international agreement will be able to be concluded.

Finally, because of the global nature of the problem, states continue to be reluctant to enter into any action unilaterally, lest others 'free ride'. Thus decision-makers want reasonably to expect that any global agreement reached will be entered into, and honoured by, all states of the world. In this way, some sort of verification process may be needed. This may well be very difficult, because greenhouse gases are produced by every state in the world and are discharged from a wide variety of sources. Given the breadth of production, increasing the respect for international agreements on global warming will be a significant challenge. Not only will reporting be required, but on-site audits and remote sensing may also be necessary. These procedures would obviously entail the collection of large amounts of physical data from all states of the world. Some national leaders would be reluctant to let others – even international organisations – undertake such a compilation, because they might suspect the motivation behind the act. Knowledge is power, and the command of information can be of strategic importance. Thus, national leaders may not let it be collected so easily. In this way, 'data sovereignty' may hamper efforts to address this issue.

In spite of these monumental challenges, politicians hope that a convention will be ready for signing by June 1992. At this time, national leaders from around the world will attend the UN Conference on Environment and Development in Rio de Janerio, Brazil. Celebrating the twentieth anniversary of the landmark UN Stockholm Conference on the Human Environment (a meeting which continues to be the single most important event in international environmental politics), the 1992 conference hopes to have, as its centrepiece, the signing of a global convention on climatic change. If this can be achieved, then the 1992 Brazil conference may be considered just as significant as its predecessor, because a tangible, worldwide response to global warming would thus be finally evident.

CONCLUSION

This chapter has attempted to provide an overview of the international dimensions of the ozone layer depletion and global warming issues. The natural processes involved are truly global – affecting all parts of the earth. Additionally, the changes that could be thus induced have the potential to affect significantly human populations, animals, material objects and the broader natural environment. The response of the international community to the challenges of global environmental change has thus far been mixed: considerable progress on the ozone layer case, but little of tangible value coming out of the global warming negotiations. Although the future is unclear, it is evident that a wide variety of factors – including the roles of science, domestic economic costs, North-South issues and compliance – will keep global environmental change upon the international political agenda.

Notes

A different version of this chapter originally appeared in *The Washington Quarterly* (Vol. 14, No. 1, Winter 1991) pp. 99–114 under the title of 'The Security Challenges of Global Environmental Change'. The author would like to thank Brad Roberts for his comments on that article. The research for this chapter was supported by grants from the Commonwealth Scholarship Commission in the United Kingdom and the Canadian Institute for International Peace and Security. The author is grateful to these organisations for their generous support.

1. EPA Report entitled *Costs and Benefits of Phasing Out Production of CFCs and Halons in the United States*, cited in 'Testimony of David D. Doniger before the Subcommittee on Health and the Environment Committee on Energy and Commerce, US House of Representatives, 25 January 1990'.
2. *Ibid.* Emphasis in original.
3. See the work of Veerhabadrham Ramanathan for information on the scientific linkage between these two issues. For example, V. Ramanathan *et al.*, 'Trace Gas Trends and Their Potential Role in Climate Change', *Journal of Geophysical Research* (Vol. 90, 1985) pp. 5547–66.
4. In 1966, Hampson suggested that significant stratospheric ozone reductions might be caused by water vapours. Crutzen (in 1970) and Johnston (in 1971) postulated that nitrogen oxides might deplete stratospheric ozone. See R. P. Turco *et al.*, 'SSTs, Nitrogen Fertiliser and Stratospheric Ozone', *Nature* (Vol. 276, 21/28 December 1978) pp. 805–7.
5. For further information on the SST-environment debate, see Joel Primack and Frank Von Hippel, 'Scientists, Politics and SST: A Critical Review', *Bulletin of the Atomic Scientists* (Vol. 28, No. 4, April 1972) pp. 24–30; and Douglas Ross, 'The Concorde Compromise: The Politics of Decision-Making', *Bulletin of the Atomic Scientists* (Vol. 34, No. 3, March 1978) pp. 46–53.
6. Nicholas Valery, 'SSTs are Clean – In Small Numbers', *New Scientist* (Vol. 68, No. 969, 2 October 1975) pp. 19–21.
7. *WMO Bulletin* (Vol. 25, No. 1, January 1976).
8. *Ibid.*
9. Mario J. Molina and F. S. Rowland, 'Stratospheric Sink for Chlorofluoromethanes: Chlorine Atom-catalysed Destruction of Ozone', *Nature* (Vol. 249, 28 June 1974) pp. 810–12. In 1973, suggestions were made that chlorine released by NASA's proposed space-shuttle might destroy stratospheric ozone. See Lydia Dotto and Harold Schiff, *The Ozone War* (Garden City, NY: Doubleday and Company, Inc., 1978) pp. 123–44. These investigations helped to set the stage for the ensuing debate.

10. See Asit K. Biswas (ed.), *The Ozone Layer* (Oxford: Permagon Press, 1979). Out of this conference, the World Meteorological Organisation (WMO) was designated the lead agency in the scientific investigations. Additionally, the 'Co-ordination Committee on the Ozone Layer' was formed to develop the Action Plan. It first met in Geneva in November 1977.
11. For a review of the EC position on the issue, see Markus Jachtenfuchs, 'The European Community and the Protection of the Ozone Layer', *Journal of Common Market Studies* (Vol. 28, No. 3, March 1990) pp. 261–77.
12. See, for example, David Dickson, 'Congress Faces Decision on CFC', *Nature* (Vol. 293, 3 September 1981) pp. 3–4.
13. David D. Doniger, 'Politics of the Ozone Layer', *Issues in Science and Technology* (Spring 1988) p. 87.
14. Thomas H. Maugh II, 'What is the Risk from Chlorofluorocarbons?', *Science* (Vol. 223, 9 March 1984) p. 1052.
15. The negotiations progressed under the name of the 'Ad Hoc Working Group of Legal and Technical Experts for the Elaboration of a Global Framework Convention for the Protection of the Ozone Layer'. See *UNEP Annual Report of the Executive Director* (Nairobi: UNEP, 1982) p. 136.
16. For more information on this conflict, see Richard Elliot Benedick, 'Ozone Diplomacy', *Issues in Science and Technology* (Vol. 6, No. 1, Fall 1989) pp. 43–50.
17. 'United Nations: Vienna Convention for the Protection of the Ozone Layer', *International Legal Materials* (Vol. 26, 1987) pp. 1516–40.
18. J. C. Farman *et al.*, 'Large Losses of Total Ozone in Antarctica Reveal CLOx/NOx Interaction', *Nature* (Vol. 315, 16 May 1985) pp. 207–10.
19. Sharon L. Roan, *Ozone Crisis: The 15-Year Evolution of a Sudden Global Emergency* (Chichester: John Wiley & Sons, Inc., 1989) p. 142.
20. *Ibid.*, pp. 189–90.
21. Tim Beardsley, 'Global Limit for CFC Emissions', *Nature* (Vol. 323, 25 September 1986) p. 286.
22. UNEP, *Action on Ozone* (Nairobi: UNEP, 1989) p. 8.
23. 'United Nations: Protocol on Substances that Deplete the Ozone Layer', *International Legal Materials* (Vol. 26, 1987) pp. 1541–61.
24. See, for example, Kathy Johnston, 'First Steps in Ozone Protection Agreed', *Nature* (Vol. 329, 17 September 1989) p. 189.
25. David Lindley, 'Ozone Hole Deeper Than Ever', *Nature* (Vol. 329, 8 October 1987) p. 473.
26. Joseph Palca, 'CFCs Phased Out', *Nature* (Vol. 332, 31 March 1988) p. 385.
27. The scientific group was called the 'Ozone Trends Panel'. It was set up in October 1986 'in response to claims made before [the US] Congress

by Donald Heath, of the NASA/Goddard Space Flight Center, that ozone levels were decreasing globally at an inexplicably high rate'. David Lindley, 'CFCs Cause Part of Global Ozone Decline', *Nature* (Vol. 332, 24 March 1988) p. 293.

28. See, for example, 'China Attacks "Unfair" Protocol', *New Scientist* (11 March 1989) p. 26; and Geoffrey Lean, 'A Welcome New Atmosphere', *The Observer (London)*, 7 May 1989.
29. CFC production plants are typically small, they can be erected quickly and they can pay for themselves swiftly as well. See Annette M. Capretta, 'The Future's So Bright, I Gotta Wear Shades: Future Impacts of the Montreal Protocol on Substances that Deplete the Ozone Layer', *Virginia Journal of International Law* (Vol. 29, 1989) p. 235, note 160.
30. 'The World Through Green-Tinted Specs', *The Economist*, 19 May 1990, p. 94.
31. Nicholas Schoon, 'Deal to Save Ozone Layer', *The Independent (London)*, 30 June 1990, p. 1.
32. *Ibid.*
33. Paul Brown, 'World Unites on Ozone Deal', *The Guardian (London)*, 30 June 1990, p. 1. Article 6 of the Montreal Protocol provides for continual assessment and review of the control measures.
34. George. M. Woodwell, 'The Effects of Global Warming', in Jeremy Leggett (ed.), *Global Warming: The Greenpeace Report* (Oxford: Oxford University Press, 1990) p. 127. See, generally, the report of Working Group II.
35. Svante Arrhenius, 'On the Influence of Carbonic Acid in the Air Upon the Temperature on the Ground', *Philosophical Magazine* (Vol. 41, April 1896) pp. 237–76.
36. Luther J. Carter, 'The Global Environment: MIT Study Looks for Danger Signs', *Science* (Vol. 169, 14 August 1970) p. 661.
37. W. W. Kellogg, 'Prediction of a Global Cooling', *Nature* (Vol. 280, 16 August 1979) p. 615. See also 'Declaration of the World Climate Conference', reprinted in *Environmental Policy and Law* (Vol. 6, 1980) pp. 103–4.
38. Stewart Boyle and John Ardill, *The Greenhouse Effect: A Practical Guide to the World's Changing Climate* (Hodder & Stoughton: New English Library, 1989) p. 41.
39. 'The Changing Atmosphere: Implications for Global Security' (Toronto, Canada, 27–30 June 1988), *Conference Statement* (Ottawa: Environment Canada, 1988).
40. Stephen H. Schneider, *Global Warming: Are We Entering the Greenhouse Century?* (San Francisco, CA: Sierra Club Books, 1989) pp. 194–8.
41. UN General Assembly, *Resolutions and Decisions Adopted by the General Assembly During its Forty-Third Session* (New York: United Nations, 1989). In 1987, the IPCC was formed by UNEP and WMO. It

met for the first time in Geneva in November 1988. It had three working groups, which examined the science, the potential socio-economic impacts and the possible response strategies. For more information, see Jack Fitzgerald, 'The Intergovernmental Panel on Climate Change: Taking the First Steps Towards a Global Response', *Southern Illinois Law Journal* (Vol. 14, Winter 1990) pp. 231–55.

42. See 'Declaration of the Hague', reprinted in *Environmental Policy and Law* (Vol. 19, No. 2, April 1989) p. 78.
43. 'G7: Economic Declaration, Section on the Environment', *Environmental Policy and Law* (Vol. 19, No. 5, 1989) p. 183.
44. 'Gandhi Urges An Environment Fund', *The International Herald Tribune*, 6 September 1989.
45. 'Global Warming Conference Falters', *The International Herald Tribune*, 7 November 1989, p. 2.
46. Michael Weisskopf, 'Bush Says More Data on Warming Needed', *The Washington Post*, 18 April 1990, pp. A1 and A23.
47. John Hunt, 'US Stand on Global Warming Attacked', *The Financial Times (London)*, 30 August 1990.
48. Nicholas Schoon, 'International Law to Protect Climate "Ready by 1992"', *The Independent (London)*, 8 November 1990, p. 6.
49. John T. Houghton (Chairman of the Scientific Assessment Working Group of the IPCC), 'Assessment of Global Warming' (Letters), *The Independent (London)*, 6 November 1990, p. 18.
50. Myles Allan, 'No Time for Dead Certs', *The Guardian (London)*, 9 November 1990, p. 29.
51. Frances Williams, 'UN Plea for Immediate Action to Curb CO_2', *The Independent (London)*, 30 October 1990, p. 2.
52. Schoon, *op. cit.*, in note 48, p. 6.
53. *Ibid.*

4 Obtaining International Environmental Protection through Epistemic Consensus

Peter M. Haas

INTRODUCTION

It is widely accepted that transboundary pollution problems require international co-operation for their solution, because many countries suffer the effects of such degradation and no country is unilaterally capable of managing the issue. While most environmental issues share these characteristics, which inhibit their resolution internationally, these issues are really little different from other global issues involving 'common property resources' – for example, trade and security. In all of these matters, national leaders are usually averse to serious co-operative efforts because of their doubts about reciprocity and verification, as well as common political antipathies among states.[1] While these problems appear to reflect some mutual interests – all countries are affected by environmental degradation and a unified response is universally preferable to a patchwork of disjointed efforts – enduring and profound differences of interest impede co-operation. Differences about who pays, states' unwillingness to forego short-term economic welfare and other distributional disagreements typically inhibit the formation of strong collective arrangements. Environmental issues are little different from the type of zero-sum bargaining efforts with which students of international relations are so familiar. Further, many less developed countries (LDCs) view efforts by the developed world to promote environmental protection as duplicitous efforts to retard economic growth in the Third World.[2] In short, mutual interests are actually quite weak, and countries are often reluctant to co-operate unless they are certain that the protection costs will be equally distributed. Further, few environmental problems are truly global in so much as they are jointly produced and consumed by all countries of the world. Most commonly discussed cases of transboundary pollution and environmental degradation occur regionally, or even sub-regionally.

Ozone layer depletion and global warming appear to be the only cases that are truly global.

How then is co-operation to occur? Recently, in the environmental sphere, it appears that a new form of co-operation has emerged. It is driven not only by state power, but also by the collective beliefs of transnationally organised networks of knowledge based communities. While it may not be completely new – Richard Cooper describes a similar process for the control of small-pox in the nineteenth century[3] – it is surely becoming more widespread with the growing reliance on technical expertise to manage highly complex international policy matters. In this article, I outline the dimensions of this incipient mode of environmental co-operation. Further, I offer a theoretical explanation for its form and some propositions about the dynamics of its reproduction. Finally, I assess its applicability for organising collective responses to the global environmental threats of ozone layer depletion and climatic change.

INTERNATIONAL ENVIRONMENTAL CO-OPERATION

There has been a striking amount of successful environmental co-operation in the past two decades: in terms of statements or commitments (the signing of international treaties), in terms of deeds (policy-making and investments) and in terms of outputs (environmental quality). For example, the International Maritime Organization (IMO) has successfully promoted a large number of international treaties regulating oil pollution from tankers, and the International Union for the Conservation of Nature and Natural Resources has developed a set of treaties which provides a framework for preserving endangered species.[4] Further, of the 140 multilateral environmental treaties that had been signed by 1988, 67 per cent were signed between 1970 and 1988.[5]

Moreover, public statements by high ranking governmental officials demonstrate a growing 'environmental consciousness'.[6] It has gone beyond mere rhetoric as expenditures and legislation in support of such efforts have increased in various countries.[7] Environmental quality has even improved in the advanced industrial societies that have undertaken such efforts. Nevertheless, environmental problems seem to be becoming more acute in industrialising LDCs, and coastal pollution from land-based pollutants remains a threat.[8]

Since its establishment in 1973, the United Nations Environment Programme (UNEP) has supported negotiations for 40 per cent of the environmental treaties signed during this period. UNEP's Regional Seas Programme sponsored 23 treaties between 1975 and 1983. The agreements covered 10 regional seas and were largely based on its experience with the Mediterranean Action Plan.[9] UNEP officials were prominent actors in these negotiations. Executive-Director Mostafa Tolba played a major role in the conclusion of the 1987 Montreal Protocol on Substances that Deplete the Ozone Layer and the 1989 Basel Convention on the Control of Transboundary Movements of Hazardous Wastes and their Disposal, and UNEP staff members were responsible for much of the success in combatting pollution in the Mediterranean.

Recent efforts under UNEP's guidance demonstrate that substantial changes in the process of international environmental treaty-making have occurred. First, there has been a shift to dealing with land-based sources of pollution, away from prior efforts involving the relatively easier and less costly problems of marine dumping and pollution. Environmental diplomacy has also become more comprehensive, covering more pollutants and more sources of pollution than previously. Further, a number of environmental treaties that are based on ecosystemic logics (for example, regional seas, river basins and airsheds), rather than political borders, have been adopted. Before UNEP's involvement, this type of treaty-making had been limited to fisheries. In addition, the 1989 signing of the Basel Convention suggests that the international community is ready to address, for the first time, the economic channels by which pollutants are transmitted. This would be in addition to the traditional manner of managing the physical channels, or environmental media, through which pollutants are transmitted.

Thus, in the aggregate and viewed over time, the evolution of environmental problem solving may be construed as a process of learning. International decision-makers have gradually come to accept a broader, interdependent, symbiotic and holistic conception of the environmental system for their pollution control efforts, and they have adapted their policies accordingly. Previous collective efforts focused largely on discrete problems that were independent of the interaction among issues. Most arrangements, however, still remain remedial rather than preventative; co-ordinating policies to regulate emissions rather than addressing the conditions that give rise to their creation.

While many of these advances in environmental co-operation were achieved under UNEP's guidance, UNEP's activities are part of a broader transnational effort. It is true that to some extent such co-operation is instigated and promoted by an international organisation. The organisation's effectiveness, however, arises from a broader set of forces of which the organisation itself is a part.

THEORY DEVELOPMENT

Unlike most instances of international co-operation, these environmental treaties were concluded without any single state – not even the United States – assuming a leadership role. I find this particularly auspicious, given the putative decline of American leadership on the global front. Co-operation without such hegemonic guidance (or control) may thus be possible.

Rather, these treaties were concluded through the influence of 'epistemic communities'. Epistemic communities are transnational networks of knowledge based communities that are both politically empowered through their claims to exercise authoritative knowledge and motivated by shared causal and principled beliefs. In the standard game-theoretic and rational-choice analyses of the international political system, epistemic communities are commonly neglected. By therefore shifting the focus to these goal seeking actors, the scholarly study of international relations is influenced in two ways: both the substantive role of ideas as a motivating source of national interest, and the question of institutional learning – as different governments respond to the provision of consensual knowledge – are reintroduced into the field.[10]

Finally, this shift in focus may be useful from a normative perspective, because epistemic communities can contribute to improved environmental policies. Their involvement in international policy-making can yield more efficient and comprehensive international policies than would be ordinarily produced. Members of these epistemic communities are less impeded by institutional rigidities and disciplinary blinders, which are traditionally encountered when complex interdisciplinary subjects are allocated to agencies that have a domain defined by one discipline and one discrete problem. Typically, such agencies generate narrow policy prescriptions which neglect the interrelationships among issues and across dimensions of analysis.

Ecological Epistemic Communities and Environmental Co-operation

Generally, international co-operation has been led by a powerful actor. Yet under recent conditions of uncertainty, specialists have played a significant role in attenuating such uncertainty for decision-makers.[11] As is often the case in environmental issues, national leaders have been uncertain about the extent of pollution, the interaction among specific pollutants, the costs of clean up and the likely actions of their neighbours. These questions are particularly puzzling in technical issues that pose low probability but high risk outcomes. Traditional search procedures and policy-making heuristics are impossible, and specific state interests may be hazy. Under such circumstances, information is at a premium, and leaders, in order to attenuate such uncertainty, may be expected to look for individuals who are able to provide authoritative advice, on whom to pin the blame for policy failure or simply as a stop-gap measure to appease public clamour for action. Politicians may also look for technical solutions which confer political gains as well. Thus, decision-makers may consult specialists for policy advice, or they may delegate responsibility to them. Subsequent discussions and policy debates will then be informed and bounded by the advice leaders receive. International negotiations may thus be viewed 'as a process for reducing uncertainty',[12] and during these negotiations, leaders defer to individuals regarded as 'experts' in the issue-area under consideration.

Traditionally, states have been impervious to technical advice. Such information was usually subordinated to shorter-term political concerns. Moreover, studies suggested that experts tended to be co-opted or defused by able politicians. Yet if the group of specialists share a common world view or believe in the same set of cause-and-effect relations, then they are likely to be particularly influential, because they will resist the political temptations to subordinate their advice to existing political concerns. When involved in decision-making, their advice will reflect the epistemic community's shared causal beliefs. In environmental matters, these views will concern the kinds of substances to be controlled, the methods to be used and the values to be employed in order to direct policy towards desired ends.

These experts, taken together, comprise an epistemic community in environmental issues. Every member of this group of specialists shares a number of beliefs: principled values regarding the enhancement of collective welfare, the validity of cause-and-effect relationships, truth tests (while most ecological epistemic communities share an acceptance of the scientific method, truth tests may in principle be based on other techniques) and a common policy enterprise. By conducting interviews and studying specialised publications of technical advisors (before their entry into policy-making), we can ascertain if these common beliefs exist and thus if an epistemic community might be present.

An epistemic community could be influential in defining the dimensions of a problem and could consequently be privileged in identifying likely solutions. As goal seeking actors, motivated by ideas and beliefs, their members will inform policy advice; they will not merely react to environmental conditions or circumstances. Their political influence will follow from their consolidation of bureaucratic power (in areas such as budgetary finances, staffing and enforcement authority). Embedded in international organisations such as UNEP, they may help set agendas, bound discussions, present issues for collective treatment and propose solutions. Embedded in national administrations, they may facilitate agreement at international negotiations, as well as guide their country's foreign policy in ways consistent with their own technical understanding and preferences. Under circumstances where foreign policy-makers may be genuinely uncertain about the identity of their interests, members of an epistemic community may prove influential in helping them identify or recognise their interests. Where interests are clearer (and uncertainty is less), they may be invited to propose policies by which desired ends may be obtained. Thus, by studying epistemic communities under settings of uncertainty, the analyst may better understand the procedure by which states' ends and means are identified, the dynamics of their decision-making process in technical issues and the rationale behind collectively formed arrangements.

Epistemic communities have been comprised of ecologists in the case of environmental issues. Ecology has been described as a framework that assimilates other scientific disciplines:

> This new science, which is an integration of the traditional disciplines of geology, oceanography, ecology, meteorology, chemistry, and other sciences, has a variety of names. . . . Its subject is nothing less than the composition, behavior and interactions of the planet's nonliving realms or phases – the atmosphere, geosphere, and hydrosphere – and its living realm, the biosphere, which encompasses parts of each of the others.[13]

Ecologists often come from diverse and hybrid disciplinary backgrounds, but they share a common belief in the need for an holistic analysis that is sensitive to the possible feedback and synergistic relationships among a variety of variables.[14] Despite uncertainty about the volume and dispersion of pollutants in the ocean, for instance, they are in agreement about the relationship between inputs of contaminants and effects on the ecosystem. Further, they aim to persuade governments to protect the environment and to encourage more self-reliant development (for example, by supporting economic and educational policies that are based on indigenous factors). Unlike many other bodies of knowledge that are regarded as being determined or conditioned by socio-economic structures, the ecological discipline, at least in its most holistic, integrative and hybrid form, does not appear to reflect and reproduce patterns of inequality and dominance that may exist in domestic and international society more broadly.[15] Ecologists' principled beliefs are focused on the need for environmental protection as an absolute end; they do not view environmental policies in terms of opportunity costs, as some economists commonly do. Consequently, when involved in international environmental negotiations, they have encouraged behaviour that is different from previous patterns of collective-action.

Propositions About Creating and Maintaining Epistemic Communities

The claim here is that if epistemic communities exist, and if they can maintain fairly stable access to decision-makers and keep rivals at bay, then international arrangements that closely resemble the community's preferences will develop and will endure. Based on these assumptions about information searching by governments and the common interests of members of an epistemic community, a number of particular propositions can be offered. In addition, techniques may be developed to improve such a process and to facilitate its application elsewhere. Thus, these propositions are not only hypotheses for research into international environmental co-operation, but they are also recommendations to guide practice that may promote such co-operation. The key to such a process, as members of UNEP have recognised, is that political arrangements and technical environmental assessment must develop concurrently.

The ten propositions are:

(1) Crises precipitate searches for new authoritative sources of policy advice. In principle, there may be an optimal frequency of crises that is sufficient to focus attention on the problem, but not so frequent as to undermine the authority of those responsible for formulating policy. Crises must be of a sufficient size to generate attention, yet not so large as to be fully terminal. If perceived uncertainty by politicians is high and public pressure is severe, then epistemic communities may be effectively able to promote policies that are further from the political 'norm' than if uncertainty was less.

(2) Epistemic communities, once identified and mobilised, may be both potent actors and the appropriate level of analysis for understanding international environmental co-operation. When organised into transnational networks, they will contribute to the adoption of concordant policies in the countries in which they gain significant bureaucratic control. Epistemic communities may also occur nationally. In such cases, their impact will be felt in the country in which they exist and in other countries that their home country is capable of influencing.

(3) Rough technical consensus established within the community will

be its principal claim to authority.

(4) Access to international and domestic authorities will contribute to influence over policy formulation and enforcement. Institutions, such as UNEP or domestic environmental ministries, can perform a number of important roles: provide financial support for ongoing research; publish scientific papers; maintain contacts within the community through, for example, conferences and telecommunications; and so on. Once established within an administration, an epistemic community is likely to retain its influence until a new crisis or paradigmatic threat challenges its authority.

(5) Scientific advice is most effective when provided through domestic channels. Technical advice from foreign scientists or international organisations is often highly suspect, especially in LDCs. The development of indigenous scientific capability reinforces the authority of those scientists providing advice to decision-makers.

(6) Consolidated bureaucratic power reinforces this influence at both the international and the national levels. International organisations are likely to be more effective at raising consciousness, circulating information, setting agendas and monitoring environmental quality. National level involvement is likely to influence the identification of national objectives, the preparation of bargaining strategies, the formulation of national positions for international negotiations and the degree of compliance with international arrangements. The creation of external advisory boards may reinforce epistemic communities' bureaucratic power by establishing sources of leverage external to their own administrative position.

(7) Expanded domestic support reinforces bureaucratic consolidation. Although environmental policy is usually elite driven, broader domestic support can reinforce the bureaucratic influence of an epistemic community within national administrations. Despite the recent emergence of Green Parties throughout Europe, actual governmental policies for environmental protection and for international co-operation preceded this mass mobilisation of public opinion. Nevertheless, public commissions, such as the World Commission on Environment and Development and the preparations for the 1992 United Nations Conference on Environment and Development, may perform a number of functions: publicise events, build up political support for continued backing of international arrangements, circulate information and elaborate domestic policies in compliance with such arrangements. International efforts may spawn

new domestic groups, or their designers may take advantage of existing networks of nongovernmental organisations (NGOs) to amplify their message. The media may also help to expand domestic support.

(8) The manner in which matters are sequenced can play an important role in the negotiations. A number of successful co-operative environmental efforts have started with umbrella assertions of fundamental premises – that is, primarily 'motherhood statements'. These basic agreements are often easier to conclude, and many can be built from existing international regimes. For example, UNEP has often utilised a 'slippery slope' strategy. This involves initially committing states to broad arrangements with few obligations, but subsequently developing more effective and constricting protocols that would have been politically unrealistic at the outset. In this manner, domestic constituencies are in place for subsequent treaty-making, and more information is probably available to bolster their positions. Agreement may be easier to reach when leaders have international validation or support for what they already wish to do domestically. An appeal to international standards may facilitate domestic application.

(9) Not all environmental problems require standards and regulation. Some cases, such as those without a sufficient understanding of the winners and losers in case of an environmental catastrophe, may be best managed through insurance funds (for example, as the IMO has designed for oil spills).

(10) This process of co-operation is reversible. If the knowledge base of the epistemic community collapses, then the epistemic community's reputation for a monopoly on uncertainty reduction will be vulnerable. However, news of the collapse of the technical consensus will not immediately reach the higher-level decision-makers who defer to such groups. Thus, due to this time-lag, the community's erosion of authority is not likely to be immediate. Nevertheless, as counter-communities emerge to challenge the causal beliefs of the epistemic community, the natural exchange of scientific ideas will gradually expand to both the broader policy world and the general public. Because the reporting of scientific matters is now more sophisticated, the rate at which information is dispersed may be fairly rapid.

One must remain cautious about the general application of such processes. They are likely to remain issue-specific, and epistemic communities will not promote political or other forms of integration. Further, the extent to which pressures may inhibit the creation of an epistemic community are not yet well understood. Moreover, epistemically guided bargaining need not always lead to co-operation. Rather, the community's advice that informs collective behaviour will simply reflect the community's causal beliefs. For instance, in the case of the United Nations Conference on Trade and Development's commodity negotiations, economists concurred that co-operation would not necessarily enhance national welfare, and subsequently the negotiations virtually collapsed.[16] Neither common interests nor even greater transparency of interests need necessarily lead to improved chances for co-operation. Whether epistemic consensus leads to co-operation depends upon the nature of the substantive consensus within the community.

ECOLOGICAL EPISTEMIC COMMUNITIES AND CO-OPERATION IN EFFORTS TO PROTECT THE OZONE LAYER

During the negotiations and implementation of the 1987 Montreal Protocol on the ozone layer, a group of ecologically minded atmospheric scientists played a major role. This epistemic community channelled discussions towards a forceful treaty by disseminating information that suggested the need for strong international controls on chlorofluorocarbons (CFCs). Further, they also helped to identify the objectives of the key country, the United States.

CFCs are industrial chemicals that are used principally for refrigeration, cooling, insulation and electronics cleaning. Invented in the early 1930s, they were widely regarded as an ideal chemical until the early 1970s, because they are inert, non-toxic, non-carcinogenic and non-flammable. In 1974, Sherwood Rowland and Mario Molina – two chemists at the University of California at Irvine – proposed a hypothesis that related CFC use to the depletion of stratospheric ozone. Subsequent widespread public fear in the United States and Europe led to the adoption of a number of regulations: a ban on the use of CFCs as aerosol propellants in the United States and Scandinavian countries and a cap on production levels in Europe.

Following this initial spate of legislation, total CFC production fell in 1977 and levelled off until 1983. In the interim, additional applications for CFCs were developed – the most significant being their use as a solvent for computer chip manufactures. Concern was renewed with a resurgence in total CFC production in 1983 and 1984. This led 20 countries to sign a framework Convention for Protection of the Ozone Layer in Vienna in March 1985. In light of the disparate policy paths pursued in Europe and North America, delegates were unable to agree on the appropriate form of regulation. Because some states wanted to limit production while others wanted to regulate different types of consumption, only a weak treaty, without targets or timetables, could be agreed upon.[17] Further, the Americans and Scandinavians had a much greater sense of urgency than most of the continental Europeans and certainly more than all of the LDCs.

However, in 1985, a new study by the British Antarctic Survey revealed a seasonally depleted ozone layer over the Antarctic, for which CFCs may have been responsible. This report triggered renewed international concern.

Committed to developing a stringent treaty, UNEP Executive-Director Tolba organised a number of scientific meetings during 1986 on questions of risk and policy responses. Negotiations on a specific protocol, replete with emission standards, resumed in December 1986. These negotiations led to the conclusion of the Montreal Protocol in September 1987, which established a schedule for a phased 50 per cent reduction in consumption and production of CFCs by 1999.

The rapid adoption of the Montreal Protocol was the consequence of extensive pressure applied by the United States at international negotiations. It is striking that during this period true consensus had not yet emerged, since there was no irrefutable evidence that identified CFCs as the cause of observed ozone depletion. Later, with growing scientific consensus, movements for enhanced controls were driven by scientific findings that incriminated CFCs.[18]

In mid-1986, a new, more stringent US position was formulated after interagency discussions between the Environmental Protection Agency's (EPA) Office of International Activities and the State Department's Bureau of Oceans, Environment, and International Science (OES) were held. After personnel changes were made by the Reagan Administration to remove officials who were appointed during the first term and who were publicly regarded as too anti-environment, both of these organisations were staffed by individuals who were sympathetic and responsive to the epistemic community. In fact, the bulk of their scientific advice came from members of this community.

The position formulated by the EPA and the OES called for a 95 per cent cut in the production of a wide variety of CFCs, to be followed by their ultimate elimination. After winning an administrative policy review battle, which lasted most of the first half of 1987, this preferred policy received the endorsement of President Reagan in June 1987.

Both the UNEP and the American positions were shaped by the views of a transnational epistemic community of atmospheric scientists. These individuals accepted the Rowland-Molina hypothesis, developed models to elaborate it and began monitoring for actual ozone depletion, while also publicly supporting policies that would ban CFCs. Further, this group of scientists shared causal beliefs that related stratospheric ozone depletion to the infusion of chlorine from CFC use. They were also bound by a belief in the utility of the scientific method for yielding 'true' and useful findings. Because of their desire to protect the environment, they advocated anticipatory action, despite the remaining scientific uncertainties. Although it was comprised mainly of atmospheric physicists and chemists from the US (where 80 to 90 per cent of the world's atmospheric science is conducted), this group had members throughout the world. Other members included the staff of UNEP and distinguished scientists in the United Kingdom and West Germany. High level officials in the US government (in the Department of State and the EPA) shared interests with this group, although they lacked the appropriate scientific training. Many of the members of the epistemic community were in frequent direct contact with one another, and they were familiar with each other's work through conferences and publications.

The US position was shaped by the beliefs and interpretations of this epistemic community. At international meetings throughout 1987, American representatives impelled other countries to accept the more stringent US position. By threatening to act unilaterally, in response to Congressional initiatives and lawsuits introduced by environmental groups, the US government had the potential to curtail access to the world's largest market for CFCs. The gradually evolving scientific consensus also reinforced the US bargaining position, although a solid consensus did not finally emerge until after the conclusion of the Montreal conference.

A major turning point in the negotiations occurred at a meeting in Wurzburg, West Germany in April 1987. UNEP Executive-Director Tolba convened a gathering of atmospheric modellers and instructed them to compare their models and assessments of ozone depletion. The resulting report provided consensual support for the position already promoted by the United States and UNEP. Given the authority of the scientists involved, this report was sufficient to compel the recalcitrant Europeans to accept the need for speedy controls over a wide variety of CFCs – a position which they had previously opposed.

While the epistemic community was thus successful in bounding discussions on the broad array of substances to be covered and the rapidity of regulations, a number of additional compromises and side-payments had to be made in order to obtain agreement. The epistemic community was responsible for identifying and bounding the broad range of alternatives, while able US diplomacy yielded the necessary compromises to sell them to other countries. A complicated formula was thus adopted, which reflected the large number of allowances necessary: there were controls on both consumption and production of CFCs (to settle US–EC differences), the cuts were to be staggered over time, the USSR was allowed to complete construction of a CFC plant, shifting the use of CFCs was allowed (at Japan's insistence) and the LDCs were given a ten year exemption from compliance. Although only covering bulk chemicals, the treaty provides for future annexes for products made with and containing CFCs. The treaty also calls for ongoing scientific reassessment to determine if stratospheric changes require modifications in the treaty's arrangements.

In the three years since the Montreal Protocol was opened for signature, increasing concern with the ozone layer has prompted significant international activity. Ozone depletion reached a historic high in 1987, and in March 1988, a new multinational Ozone Trends Report, which presented the findings from Antarctica and thus related ozone depletion to CFC use, was released in the US. The report triggered both extensive alarm and calls for more stringent additions to the Protocol. Du Pont, the world's principal producer of CFCs, responded within a week by announcing its intention to discontinue CFC production by the end of the century. In September 1988, EPA Administrator Lee Thomas called for the complete elimination of CFCs. Two months later, the British government announced that it was in favour of 85 per cent cuts. In February 1989, evidence from Antarctic studies demonstrated that ozone was again being depleted more rapidly than originally predicted. In early March 1989, EC Environment Ministers called for full elimination of CFCs as soon as possible, and in the following week, 123 countries called for absolute elimination of the chemicals by the end of the century. In May 1989, at the first governmental review meeting of the Montreal Protocol, 81 countries adopted a resolution calling for a total phase out of all CFC production and consumption by the year 2000. Delegates also called for a fund to be established in order to ameliorate the costs of adjustment for LDCs. In June 1990, 93 states amended the Protocol in a number of ways: they agreed to ban the regulated CFCs by 2000, they agreed to expand the coverage to a variety of other chemicals that destroy stratospheric ozone and they agreed to establish a US$240 million international fund to ease the costs of compliance by LDCs.

This parade of environmental concern was orchestrated to a large extent by the epistemic community. Upon becoming influential in the governments of major states, the members of the epistemic community encouraged their countries to support a strong international treaty and to press later for its strengthening and rapid enforcement. In the United States, the community's entrenchment dated from 1985, with the hiring of new officials in the EPA and the State Department; in other countries, it happened later, and correspondingly, support occurred later. In West Germany, a new Environment Ministry, which was responsive to the advice and pressures of the Greens and national atmospheric scientists, was created in 1985. Support in the UK did not develop until Prime Minister Thatcher's speech to the Royal Society in September 1988, which called for immediate action to preserve the ozone layer. This turnaround was prompted by a number of factors: the submission of a new report by the British Stratospheric Ozone Review Group, which called for such action; similar advice from her United Nations Permanent Representative, who was personally concerned about atmospheric problems; and a surge in domestic environmental concern.[19]

Other, less politically influential countries that were not major producers or consumers of CFCs merely followed the lead of the states that were responsible for the majority of the consumption and production of CFCs. Thus, the ozone layer regime, created by an ecological epistemic community within UNEP and many of the industrialised countries, was accepted by other countries; their immediate economic interests as producers or consumers were less evident. Without the presence of the epistemic community, it is likely that there would have been less co-operation, that its form would have been less comprehensive (that is, fewer pollutants covered and less sensitivity to related issues), that its enforcement would have been slower and less aggressive and that the variation among national regulatory efforts would have been much broader.

ASSESSMENT OF PROPOSITIONS AND APPLICATION TO GLOBAL WARMING

The ozone case confirms many of the propositions about epistemic communities: policy action was precipitated by a crisis; a transnational epistemic community bounded the range of policy alternatives around which diplomats bargained; and, once members of the epistemic community consolidated their positions within national governments, these countries supported the international arrangements.

Experts appear to have played a widespread role in other international environmental negotiations as well. As a consequence of stressing the linkages among ecosystems, they encouraged governments both to control a broad assortment of pollutants and to redesign decision-making processes in order to integrate environmental considerations into more traditional policy-making. For example, ecologists played an important role in establishing and maintaining the Mediterranean Action Plan, other international regional seas arrangements and the diffusion of national species conservation strategies. Regularised negotiations established channels of interaction among the relevant 'technical elites' in participating countries, NGOs and international organisations.[20]

A focus on epistemic communities may also be appropriate for generating and understanding co-operation on other transboundary environmental issues. Regional and global environmental problems, such as acid precipitation and deforestation, may be resolved through the application of such a technique, because consensus exists for each issue and members of the epistemic community have been mobilised in national and international fora. Efforts to control European acid precipitation appear to have already evolved by epistemic consensus.

Yet its applicability to global warming is much less clear. Efforts are currently underway to develop a treaty to be approved by the 1992 UN Conference on Environment and Development. While the form of the treaty remains unclear, multilateral efforts have been continuing in order to find global consensus and to modify the extensive scientific uncertainties regarding the timing, the consequences and the dynamics of global climate change. Diplomats remain split about the form that collective regulations should take – a framework treaty followed by protocols regulating specific greenhouse gases or a comprehensive 'law of the atmosphere' concurrently covering a number of greenhouse gases or some form of concerted energy policies.[21] The United States has been recalcitrant in these negotiations, stressing the magnitude of the scientific uncertainties that remain. At a ministerial conference in Noordwijk, the Netherlands in November 1989, the US was marginalised from most European delegations by its reticence to acknowledge the need for international efforts to control greenhouse gas emissions. In April 1990, the Bush Administration was excoriated by the Europeans for its continued emphasis on the remaining uncertainties and for stressing increased research rather than policy action.

An incipient epistemic community may exist for climate change. At two early scientific conferences, the scientific community argued that global warming was real and that immediate policies to reduce emissions of greenhouse gases were required. The 1985 Villach Conference and the 1988 Toronto World Conference on the Changing Atmosphere put climate change on the international agenda and focused international concern on the problem. The Toronto Declaration called for a 20 per cent reduction of carbon dioxide emissions by the year 2005.

The Intergovernmental Panel on Climate Change (IPCC) was subsequently established by UNEP and the World Meteorological Organization in November 1988 to co-ordinate and integrate worldwide understanding of the scientific, economic and policy issues relating to climate change. Multinational working groups on science (chaired by the UK), impacts (chaired by the USSR) and response strategies (chaired by the US) submitted their reports at the end of 1990. These reports are providing the basis for treaty writing efforts.

While an emergent form of an epistemic community exists in the global warming issue, its consensus is nowhere near as strong as in the ozone layer issue. Although atmospheric scientists speak of their certainty that some climate change is unavoidable from the introduction of greenhouse gases, they are much less sure about the timing, extent or distribution of the effects.[22] The community is also split between preventionists (who believe in immediate action in order to prevent or delay climate change) and adaptationists (who argue for concerted responses to climate change).[23]

Further, the community's avenue to decision-making in some of the major actors is blocked. In the United States, atmospheric scientists who actively believe in the need for prompt carbon dioxide controls have been barred access to the administration. Congressional testimony by James Hansen, Director of the National Aeronautics and Space Administration's Goddard Institute for Space Studies, was changed to weaken his conclusions about global warming.[24] Strong controls are opposed by the domestically powerful Chief of Staff, John Sununu, and Director of the Office of Management and Budget, Richard Darman. Both individuals are sceptical of the high costs that such measures would be likely to invoke, and they are also averse to the implicit need for federal regulation that these measures would require.[25] Conversely, in Germany and the United Kingdom, where Green movements have become more influential, the administrations have recently become much more receptive to possible regulatory action.

It is much more difficult to reach collective agreement on climate change than ozone layer depletion, primarily because the costs of action are much higher. Achieving co-operation on the ozone layer issue was relatively easy, because the adjustment costs were ultimately borne by a small number of chemical producers that did not rely on CFCs for the bulk of their profits. Further, alternative chemicals were thought to be available – in short, a technical fix was possible. Moreover, because CFCs were also viewed as a greenhouse gas, the ozone layer issue, at least by the late 1980s, was seen as a relatively low-cost, high political gain issue to promote. Effective management of carbon dioxide and other greenhouse gases, on the other hand, would require a substantial transformation of our modern industrial and consumer life, because it has been built upon the easy provision of relatively inexpensive fossil fuels. Moreover, the costs of prevention may be higher than the costs incurred from little action.[26] In the face of such high costs, technical advice and atmospheric scientists' claims of authority are surely viewed with much greater scepticism by policy-makers.[27]

Moreover, the distribution of costs among countries, of both action and inaction, is diffused, thus making agreement on specific control measures and targets extremely difficult. The array of countries that will be significantly harmed by such actions, unless compensated, is quite large.[28] China is a vivid example of a country that would have large adjustment costs – it relies on coal for 80 per cent of its energy, and it is investing heavily in coal fired power plants for future energy generation. It now also appears that some countries and regions – because of increased rainfall for crops and extended growing seasons – perceive themselves to be potential 'winners' in the short- to medium-term. Thus, they may oppose collective measures.

Finally, the agenda is much more complex than in the ozone layer case. While the ozone layer issue was only loosely coupled with other international issues, global climate change is tightly coupled with other international concerns, because widespread socio-economic consequences would result from extensive regulatory policies to control climate change. Thus, negotiations may be much more complex, and the direct costs and opportunity costs of action may be much higher and more difficult to determine. Agreements reached by making tactical linkages between sub-issues may fail, because congruent substantive linkages deny their feasibility. Thus, while the Law of the Sea experience may be a better analogy for climate change negotiations than the ozone layer experience, the multiple tactical linkages among various sub-issues which facilitated the Law of the Sea agreement (and also the ozone layer agreement) may fail in the climate change issue.[29]

To some extent a window of opportunity of ignorance – what Oran Young calls the 'veil of uncertainty'[30] – was missed in the mid-1980s. Before evidence of winners and losers became apparent, an insurance regime would have been possible, as all countries were concerned that they would lose from global warming in the short- to medium-term. However, the countries with the most to gain now from global warming are those with the most resources to help others, while affluent losers can channel their resources to ameliorate purely domestic effects.

Collective-action on greenhouse gases is not impossible, I am simply arguing that it is unlikely to occur through epistemic consensus. By late

1990, domestic pressures had driven seven European countries to adopt policies to reduce carbon dioxide emissions. However, allowing domestic forces to prevail will make international co-ordination more difficult, because countries will have already embarked on disparate policy paths, which may prove difficult to co-ordinate.

CONCLUSION

This chapter has presented a broad process by which international environmental co-operation is possible. It has also laid out some guidelines for its study and application. In the efforts to protect stratospheric ozone, an ecological epistemic community was influential; it identified the broad scope of international policy, and it pressured governments to comply with international standards. Foreign policy-makers were uncertain about the dimensions of these problems, and they reacted to crises and demands for action by soliciting technical advice. The concept's applicability to global warming was then examined. It was noted that there are important differences between the ozone layer depletion and climate change issues. In the global warming case, there are economic and political conditions that are inhibiting the influence, authority and even the sheer existence of a coherent epistemic community. Thus, while international co-operation may be possible to limit global warming, it is unlikely to occur by the same process by which co-operation and environmental protection has been accomplished in many other environmental issues.

Notes

Earlier versions of this chapter were presented at: the International Congress Experts are Categorical: Scientific Controversies and Political Decisions Concerning the Environment, Arc-et-Senans, France, 11–13 September 1989; and the Conference on Technology-Based Confidence-Building: Energy and Environment, St. John's College Conference Center, Santa Fe, NM, 9–14 July 1989. The author is grateful to Ernst B. Haas, M. J. Peterson, Steve Rayner, Peter H. Sand and Oran R. Young for their comments on earlier versions of this chapter.

1. Oran R. Young, 'Politics of International Regime Formation', *International Organization* (Vol. 43, No. 3, Summer 1989) pp. 349–76; Oran R. Young, *International Cooperation: Building Regimes for Natural Resources and the Environment* (Ithaca, NY: Cornell University Press,

1989); Peter M. Haas, *Saving the Mediterranean: The Politics of International Environmental Cooperation* (New York: Columbia University Press, 1990); Eugene B. Skolnikoff, 'The Policy Gridlock on Global Warming', *Foreign Policy* (No. 79, Summer 1990) pp. 77–93; and Peter H. Sand, *Lessons Learned in Global Environmental Governance* (Washington, DC: World Resources Institute, 1990).

2. Peter Usher, 'Climate Change and the Developing World', *Southern Illinois University Law Journal* (Vol. 14, 1990) pp. 257–64.
3. Richard N. Cooper, 'International Cooperation in Public Health as a Prologue to Macroeconomic Cooperation', in Richard N. Cooper, Barry Eichengreen, C. Randall Henning, Gerald Holtham and Robert D. Putnam (eds), *Can Nations Agree? Issues in International Economic Cooperation* (Washington, DC: The Brookings Institution, 1989) pp. 178–254.
4. The actual effectiveness of the arrangements created by these two organisations continues to be the subject of much debate. See Sand, *op. cit.*, in note 2; Mark W. Zacher and R. Michael M'Gonigle, *Pollution, Politics and International Law* (Berkeley, CA: University of California Press, 1979); and Simon Lyster, *International Wildlife Law* (Cambridge: Grotius Publications, 1985).
5. UNEP, *Register of International Treaties and Other Agreements in the Field of the Environment* (Nairobi: UNEP, May 1989).
6. Mostafa K. Tolba (ed.), *Evolving Environmental Perceptions* (London: Butterworths, 1988); and Robert Goodland and George Ledec, 'Environmental Management in Sustainable Economic Development', in W.C. Baum and S. Tolbert (eds), *Investing in Development: Lessons of World Bank Experience* (Oxford: Oxford University Press, 1985).
7. OECD, *The State of the Environment* (Paris: OECD, 1985); and Peter M. Haas, 'Making Progress In International Environmental Protection', in Emanuel Adler and Beverly Crawford (eds), *Progress in Postwar International Relations* (New York: Columbia University Press, forthcoming).
8. UNEP, *The State of the World Environment 1989* (Nairobi: UNEP, 1989); UNEP, *The State of the World Environment 1987* (Nairobi: UNEP, 1987); Global Environment Monitoring System, *Assessment of Chemical Contaminants in Food* (London: UNEP, FAO and WHO, 1988); Global Environment Monitoring System, *Assessment of Urban Air Quality* (London: UNEP and WHO, 1988); and Global Environment Monitoring System, *Assessment of Freshwater Quality* (UNEP and WHO, 1988). See also World Resources Institute, *World Resources 1990–91* (Oxford: Oxford University Press, 1990).
9. Peter H. Sand, *Marine Environmental Law in the United Nations Environment Programme* (London: Tycooly Publishing, 1988) p. ix. See also Haas, *op. cit.*, in note 1; and Peter M. Haas, 'Save the Seas: UNEP's Regional Seas Programme and the Coordination of Regional Pollution Control Efforts', in Elisabeth Mann Borgese *et al.*,

(eds), *Ocean Yearbook 9* (Chicago, IL: University of Chicago Press, forthcoming).

10. Several alternative models of learning already exist. See John D. Steinbruner, *The Cybernetic Theory of Decision* (Princeton, NJ: Princeton University Press, 1974); Hugh Heclo, *Modern Social Politics in Britain and Sweden* (New Haven, CT: Yale University Press, 1974); Ernst B. Haas, 'Why Collaborate? Issue-Linkage and International Regimes', *World Politics* (Vol. 32, No. 3, April 1980) pp. 357–405; Ernst B. Haas, *When Knowledge is Power: Three Models of Change in International Organizations* (Berkeley, CA: University of California Press, 1989); and Haas, *op. cit*., in note 1, pp. 58–63.
11. Uncertainty in this regard is used to mean that policy-makers are unable to assess the likelihood of an event's occurrence and hence cannot apply some decision models or heuristics to estimate costs and benefits of alternative lines of action. It implies much more than merely incomplete information. Scientists and members of an epistemic community are less uncertain and better trained at coping with such uncertainty.
12. Gilbert R. Winham, 'Negotiation as a Management Process', *World Politics* (Vol. 30, No. 1, October 1977) p. 96.
13. World Resources Institute, *World Resources 1987* (New York: Basic Books, 1987) p. 163.
14. For a review of ecological thought, see Donald Worster, *Nature's Economy: A History of Ecological Ideas* (Cambridge: Cambridge University Press, 1977); Edward J. Kormondy and J. Frank McCormick (eds), *Handbook of Contemporary Developments in World Ecology* (Westport, CT: Greenwood Press, 1981); and Robert P. McIntosh, *The Background of Ecology* (Cambridge: Cambridge University Press, 1985).
15. Erich Jantsch and Conrad Waddington (eds), *Evolution and Consciousness* (Reading, MA: Addison-Wesley Publishing Co, 1976).
16. Robert L. Rothstein, 'Consensual Knowledge and International Collaboration', *International Organization* (Vol. 38, No. 4, Autumn, 1984) pp. 733–62.
17. Peter H. Sand, 'Protecting the Ozone Layer', *Environment* (Vol. 27, No. 5, June 1985).
18. The following account relies on Richard Elliot Benedick, *Ozone Diplomacy* (Cambridge, MA: Harvard University Press, 1991); Peter Morrisette, 'The Evolution of Policy Responses to Stratospheric Ozone Depletion', *Natural Resources Journal* (Vol. 29, No. 3, Summer 1989) pp. 793–820; and Peter M. Haas, 'Ozone Alone, No CFCs: Ecological Epistemic Communities and the Protection of Stratospheric Ozone', *International Organization* (forthcoming).
19. United Kingdom Stratospheric Ozone Review Group, *Stratospheric Ozone 1988* (London: Her Majesty's Stationery Office, 1988); House of Lords Select Committee on the European Communities, *The Ozone Layer – Implementing the Montreal Protocol with Evidence, 1987–88,*

17th Report, HL Paper 94 (London: Her Majesty's Stationery Office, 1988); and 'The Greening of Margaret Thatcher', *The Economist* (11 March 1989) pp. 55–6.

20. D. M. Johnston, 'Marine Pollution Agreements: Successes and Problems', in J. E. Carroll (ed.), *International Environmental Diplomacy* (Cambridge: Cambridge University Press, 1988) p. 204. The International Union for the Conservation of Nature and Natural Resources' (IUCN) World Conservation Strategy has been responsible for the creation of over 30 national groups. See Mark Halle and Jose Furtado, 'The Role of National Conservation Strategies in Attaining the Objectives of the World Conservation Strategy', in Peter Jacobs and David A. Munro (eds), *Conservation With Equity: Strategies for Sustainable Development* (Gland, Switzerland: IUCN, 1987).
21. Mostafa Tolba, 'A Step-by-Step Approach to Protection of the Atmosphere', *International Environmental Affairs* (Vol. 1, No. 4) pp. 304–9; William A. Nitze, 'A Proposed Structure for an International Convention on Climate Change', *Science* (10 August 1990) pp. 607–8; and Fen Osler Hampson, 'Climate Change: Building International Coalitions of the Like-minded', *International Journal* (Vol. 45, No. 1, Winter 1989–90) pp. 36–74.
22. Norman J. Rosenburg, William E. Easterling III, Pierre R. Crosson and Joel Darmstadter (eds), *Greenhouse Warming: Abatement and Adaptation* (Washington, DC: Resources for the Future, 1989). Atmospheric scientists acknowledged that the present state of knowledge is insufficient to demonstrate the extent of the problem or to generate credible policy relevant advice. They estimated that it would take from 0 to 5 years to produce research that would generate consensus on warming, 5 to 20 years on sea level change and over 10 years for many other aspects of the problem. See also Jill Jager, 'Anticipating Climatic Change', *Environment* (Vol. 30, No. 7, September 1988).
23. Michael H. Glantz, 'Politics and the Air Around Us: International Policy Action on Atmospheric Pollution by Trace Gases', in Michael Glantz (ed.), *Forecasting by Analogy: Societal Responses to Regional Climatic Change* (Boulder, CO: Westview Press, 1988) pp. 41–72.
24. Philip Shabecoff, 'Scientist Says Budget Office Altered His Testimony', *The New York Times*, 8 May 1989, p. A1; and Philip Shabecoff, 'White House Admits Censoring Testimony', *The New York Times*, 9 May 1989, p. C1.
25. Lynton K. Caldwell, 'International Environmental Politics: America's Response to Global Imperatives', in Norman J. Vig and Michael E. Kraft (eds), *Environmental Policy in the 1990s* (Washington, DC: CQ Press, 1990) pp. 315–16.
26. 'Count Before You Leap', *The Economist* (7 July 1990) pp. 21–4.
27. It is interesting to note that an economic epistemic community might have much greater influence, because it would be able to mitigate uncertainty about the costs of action.

28. George Rathjens, 'Energy and Climate Change', in Jessica Tuchman Mathews (ed.), *Preserving the Global Environment* (New York: W.W. Norton & Co, forthcoming).
29. See James K. Sebenius, 'Negotiation Arithmetic: Adding and Subtracting Issues and Parties', *International Organization* (Vol. 37, No. 2, Spring 1983) pp. 281–316.
30. See Oran R. Young, 'Global Environmental Change and International Governance', in Chapter 2.

5 Economics and the Global Environmental Challenge

David Pearce

INTRODUCTION: GLOBAL ENVIRONMENTAL MORALITY

For many people, the idea of promoting the greatest good is a moral intuition, something that cannot rationally be defended. The idea of caring for future generations and other sentient beings is perhaps in the same category. The environmental challenges of the last part of the twentieth century raise all the problems of 'future care', because so many of the costs of environmental degradation will be borne by future generations and by the natural world. In some recent work, my colleagues and I *assumed* the imperative of future concern; the source of the imperative was not discussed.[1] Interest centred on what such a moral principle would mean in practice. Following, but also departing from, the important writings of Solow, Page, Hartwick and Maler,[2] it was argued that concern for the future could be made operational by a rule that required each generation to pass on to the next one a stock of natural environmental assets ('natural capital') no less than the stock of assets already in existence. Simply put, we should not degrade our environment any further – we should not 'live off our capital'. By so doing, current generations could do what is feasible to compensate future generations for damage now being done; the costs of which would be largely borne in the future. This is the 'intergenerational externality' phenomenon, and the correction of this externality is required if intergenerational fairness is to be observed. In turn, intergenerational fairness is a critical constituent part of any definition of 'sustainable development'.[3]

This 'constant natural capital' rule is a variant of one that might have wider appeal to economic orthodoxy, namely, passing on a stock of natural *and* human-made capital. But the distinction between the two rules is fundamental. Leaving a constant stock of all kinds of capital is consistent with burning the Amazon forests and converting them to industrial forestry or agriculture. It would always be acceptable to run down nature's wealth provided we built up another form of wealth – roads or machinery, knowledge or technology. The arguments for narrowing the focus to nature's capital are not developed here – they are spelled out elsewhere[4] – but the rationale includes:

(1) The widespread public view that we have gone far enough, and often too far, in destroying natural environments;

(2) The high economic values that are often revealed when conservation is properly accounted for;

(3) The asymmetry between the types of capital, more specifically, human-made capital is reproducible from knowledge, while nature's is not. Human-made assets can be created or destroyed almost at will. However, once nature's capital is extinguished, much of it cannot be reproduced. Thus, there is *irreversibility*;

(4) The *uniqueness*, and hence non-substitutability, of much environmental capital – for example, the ozone layer and tropical rainforests;

(5) The extensive *uncertainty* that surrounds our knowledge of how natural ecosystems function and serve as life-support agencies.

Conserving nature's capital is an instrumental rule for being fair to future generations. It is a means of implementing the moral intuition that we should serve the greatest good, where the good includes that of future generations, sentient non-humans and current generations – and especially the poor of the current generation who tend to suffer most when environments are degraded (for example, the resource-dependent people of the Sahel or Nepal).

The constant capital concept needs a great deal of further thought.[5] But even at the superficial level, it appears inconsistent with a utilitarian approach, because it implies that reductions in natural capital cannot be contemplated regardless of the benefit of such reductions. Utilitarianism would compare the costs and benefits of actions and is thus consistent with degrading the environment provided the benefits of so doing exceed the costs. But even a utilitarian approach must account for the three features that tend to characterise modern global environmental challenges: uncertainty, irreversibility and uniqueness. Taken together, these features of the problem justify a very cautious approach to environmental capital, because the payoff from present destructive action could have extremely large negative consequences in the future. This is very much how the challenge of global warming is being characterised. The 'constant capital'

TABLE 5.1 *Summary of Key Human-Made Greenhouse Gases*

	Carbon Dioxide	*Methane*	*CFC-11*	*CFC-12*	*Nitrous Oxide*
Atmospheric Concentration	ppmv	ppmv	pptv	pptv	ppbv
Pre-industrial	280	0.79	0	0	288
Present Day	354	1.717	280	484	310
Rate of Change per annum	1.6 (0.5%)	0.015 (0.9%)	10 (4.0%)	17 (4.0%)	0.8 (0.25%)
Atmospheric Lifetime (in years)	50–200	10	65	130	150

SOURCE: Intergovernmental Panel on Climate Change.

NOTES: ppmv, parts per million by volume
ppbv, parts per billion by volume
pptv, parts per trillion by volume

approach implies high (but not infinite) values for the environment, but it does not require that as each tree is cut down, another identical one is replaced. It countenances a commitment to overall environmental conservation, not a commitment to conserve each environmental asset. In its most dilute form, the principle requires that we pay special attention to the conservation of 'critical' environmental capital – capital on which we depend for life support and spiritual well-being.

CONSERVING CAPITAL AND GLOBAL ENVIRONMENTAL CHANGE

The instrumental principle of conserving natural capital has secured a place in international negotiation on environmental change, although it is not necessarily articulated in this way. We would expect a constant capital rule to be applied most urgently to those natural resources that have the features of non-substitutability and irreversibility, and that are under the gravest threat. The resources that have these features and that are the subject of actual, imminent or near-term global agreements are the ozone layer, the atmosphere and the last great wilderness – Antarctica. The conspicuous resource that is non-substitutable and that is suffering effective irreversible damage, but that is not yet the subject of an international agreement, is the tropical forest.

In the ozone layer case, the world has agreed to what is effectively a zero depletion policy, which is formally identical with constant natural capital. In the case of greenhouse gases and climate change, discussions are heading that way.

The Ozone Layer

International action on the ozone layer was taken with the drawing up of the United Nations Convention on the Protection of the Ozone Layer in Vienna in 1985. The Montreal Protocol on Substances that Deplete the Ozone Layer, which came into force at the start of 1989, is the actual agreement to reduce the use of chlorofluorocarbons (CFCs). Under the initial agreement, *consumption* of five CFCs (11, 12, 113, 114, 115) was frozen at the 1986 levels by 1989 and had to be cut by 20 per cent of the 1986 levels by 1994, eventually to be lowered to 50 per cent by 1998–99. The cuts in

TABLE 5.2 *Greenhouse Gas Contributions to Global Warming (1990 emissions allowing for cumulative effect)*

	Global Warming Potential (100-year horizon)	*1990 Emissions (Tg)*	*Relative Contribution over 100 yrs %* (per cent)
Carbon dioxide	1	26 000	61
Methane	21	300	15
Nitrous oxide	290	6	4
CFC-11	3 500	0.3	2
CFC-12	7 300	0.4	7
HCFC-22	1 500	0.1	0.4
Other			10.6

SOURCE: Intergovernmental Panel on Climate Change.

NOTE: The relative contributions vary according to the period over which cumulation is considered. Thus carbon dioxide is typically shown as contributing around 50 per cent of warming based on 'instantaneous' impacts. The figures in this table allow for cumulative effects over 100 years – that is, they are adjusted to account for the relative lifetimes of the gases.

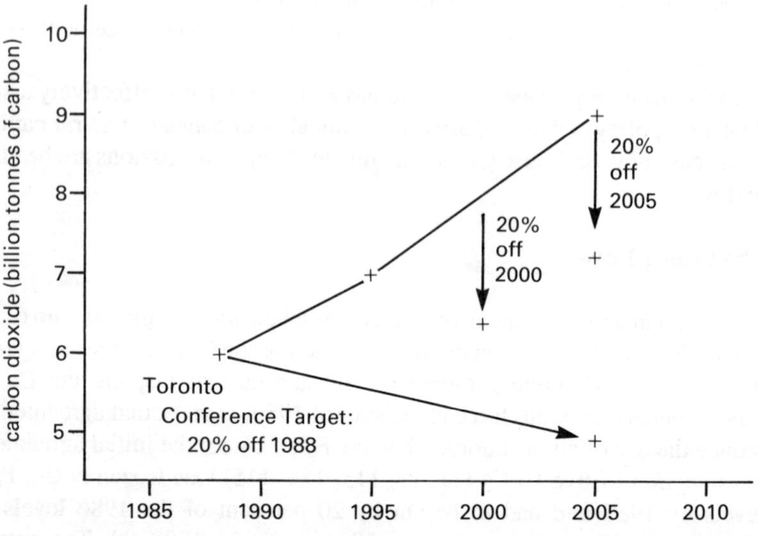

Source: Emissions estimates from International Energy Agency, *Greenhouse Gas Emissions: the Energy Dimension* (Paris: International Energy Agency, 1990).

Figure 5.1 Carbon dioxide reduction targets

production were slightly less severe in order to allow for the needs of the developing countries, with the aim being to allow 65 per cent of the 1986 level by 1999. Three halons (1211, 1301, 2402) were the subject of a 1986–based consumption freeze and a production level freeze of 110 per cent of the 1986 level by 1992. An agreement reached in London in June 1990 produced stricter controls still and revised the Montreal Protocol to achieve 50 per cent cuts by 1995 and 85 per cent by 1997. Some countries pressed for total phase-out by 1997. Halons will be cut by 50 per cent by 1995 and will be phased out by the year 2000. Carbon tetrachloride and methyl chloroform were added to the agreement – the former to be phased out by 2000, the latter by 2005. Developing countries have to achieve these targets within a grace period of ten years. Virtually all developed countries are now party to the Montreal Protocol, and these countries account for about 80 per cent of the global consumption of the relevant CFCs and halons. While 23 developing nations have signed the Protocol, India and China have not signed, but have announced their decision to do so in 1992. The London meeting also established a special fund of US$240 million to help developing countries adjust to the CFC phase-out.

Effectively then, the Montreal agreement and its modifications seek a virtual cessation of production of ozone-depleting CFCs. Although this will still result in increased chlorine releases in the atmosphere, due to the time lags between CFC production, consumption and release to the atmosphere, the Protocol is as near as we can get to the intent of treating the ozone layer in terms of the instrumental rule to maintain constant natural capital. 'Intent' is the operative word, because there are potentially formidable implications of the Protocol if developing countries do not accede and major implications even if they do accede, but increase their consumption of CFCs to the limit allowed under Article 5.[6]

Greenhouse Gases

CFCs are also greenhouse gases and are thus implicated in global warming. The other greenhouse gases are carbon dioxide, methane, nitrous oxides and tropospheric ozone arising from reactions involving methane, carbon monoxide, nitrogen oxides and sunshine.[7] Table 5.1 summarises the characteristics of the key human-made greenhouse gases, and Table 5.2 shows the relative contributions of each gas to global warming.

SETTING TARGETS: EXISTING DISCUSSIONS

We do not yet have a convention on global warming, but national positions are being determined. The US has engaged in some retraction of earlier proactive positions on global warming, primarily because of concerns over the economic costs of emissions reductions and perhaps because of some scientific reservations.[8] Japan has also expressed doubts about the wisdom of positive action, but has set an initial modest target of constant *per capita* emissions by 2000 (at 1990 levels). In contrast, European countries, notably Germany and the Netherlands, are urging strong immediate action. Indeed, they have effectively adopted unilateral targets for carbon dioxide reduction: Germany is discussing a 25 per cent reduction of 1986 emissions by 2000 and the Netherlands endeavouring for constant 1990 emissions by 1995.

In November 1989, ministers from developed and developing countries met in Noordwijk, the Netherlands and agreed the following declaration:

> [The conference] recognizes the need to stabilize, while ensuring stable development of the world economy, CO_2 [carbon dioxide] emissions

> and emissions of other greenhouse gases not controlled by the Montreal Protocol. Industrialized nations agree that such stabilization should be achieved by them as soon as possible, at levels to be considered by the IPCC [Intergovernmental Panel on Climate Change] and the Second World Climate Conference of November 1990. In the view of many industrialized nations such stabilization of CO_2 emissions should be achieved as a first step at the latest by the year 2000.
>
> [The conference] urges all industrialized countries to support the process of IPCC through the investigation of the feasibility of achieving targets to limit or reduce CO_2 emissions including, for example, a 20 per cent reduction of CO_2 emission levels by the year 2005 as recommended by the scientific world conference on the Changing Atmosphere in Toronto 1988.[9]

Representatives from 67 countries, including the US, the UK, China and India, agreed to this statement. The targets for greenhouse gas emissions are loosely worded, especially in respect of the omission of a base year for the 20 per cent reduction of carbon dioxide. However, the 1988 Toronto Conference recommendation referred to makes it quite clear that the base year for the 20 per cent reduction is the year of the conference, 1988.[10] That is, by 2005, the industrialised nations should be emitting 20 per cent less carbon dioxide than in 1988. Moreover, the Toronto conference indicated that this was only a first step: the 20 per cent has to be a global goal, not one just for the developed world, and 50 per cent cuts from 1988 carbon dioxide emission levels would be necessary to stabilise atmospheric concentrations. The State of Victoria, Australia, has officially adopted the target in its own greenhouse gas policy statement.

The Noordwijk Declaration is thus ambiguous. It could mean one of three things:

(1) Stabilising carbon dioxide emissions at their level in 2000;

(2) Securing a 20 per cent cut on what the levels would otherwise be in 2005. Figure 5.1, based on past trends in carbon dioxide emissions, shows that this is roughly equal to stabilising emissions *at current levels*;

(3) Securing a 20 per cent cut on 1988 levels by 2005.

Each of these targets may be further varied according to what grace periods might be given to the developing world. Figure 5.1 shows that the three alternative interpretations are very different. Discussion has tended to centre on stabilisation at current levels of emissions. Once again, such

TABLE 5.3 *Costs and Benefits of Greenhouse Gas Reductions*

Reductions in Greenhouse Gas Emissions as a Percentage of Base Level	*Total Bost of Reduction (US$ billion)*	*Total Benefit of Reduction (US$ billion)*
0	0	0
5	0.4	5.9
10	1.6	11.8
15	4.1	17.7
17	**5.6**	**19.6**
20	11.9	23.6
30	46.7	35.4
40	107.0	47.2
50	200.6	59.0

SOURCE: W. Nordhaus, *To Slow or Not to Slow: The Economics of the Greenhouse Effect* (New Haven, CT: Yale University, Department of Economics, mimeo., 1990). The benefit figures are from Nordhaus's 'medium damage' scenario, corresponding to damage of some $13 per tonne carbon dioxide equivalent (in 1989 US dollars).

targets are consistent with the idea of treating the atmosphere as natural capital and seeking to hold it broadly constant.

SETTING TARGETS: THE ACCEPTABLE LEVEL OF GLOBAL WARMING

The world is already committed to some global warming because of the thermal inertia of the oceans. Even if the emission of all greenhouse gases stopped tomorrow, some global warming would still occur. The Toronto Conference target of 20 per cent reduction in emissions from current levels was thought to be consistent with a rate of warming of 0.1°C each decade.[11] This rate has been advanced as an 'ecologically manageable' rate of increase in warming – that is, it is the rate at which non-disruptive ecological change will take place. This idea of using 'ecological limits' is very much akin to the constant natural capital approach.[12] Within policy circles, this rate of change of warming is increasingly being regarded as an 'acceptable' level.

There is, however, an alternative way of deriving the acceptable level of warming. This would be to approach the issue from the standpoint of comparing costs and benefits in monetary terms – that is, by adopting a utilitarian approach in which sacrifices now are weighed against future gains.[13] One significant attempt to compute 'ballpark' figures for costs and benefits has been undertaken by Nordhaus.[14]

Nordhaus suggests that the cost and benefit picture appears as in Table 5.3. The table should be read so as to find the greatest *difference* between costs and benefits – that is, maximum net benefits from reduced emissions (and from afforestation as a means of 'fixing' carbon dioxide). Thus, a 5 per cent reduction in emissions is clearly worthwhile, because the world would gain about US$6 billion in avoided damages at a cost of only US$400 million. Reductions are worthwhile up to a maximum net benefit level corresponding to 17 per cent reductions, but beyond this the net benefit figure is reduced. For cuts of approximately 30 per cent, costs actually exceed benefits.

Nordhaus shows that this overall reduction should comprise a virtual phase-out of CFCs and a reduction of 6 per cent in carbon dioxide emissions. The finding that CFCs should be phased out is wholly consistent with the existing international discussions, but this is the only correspondence. A probable interpretation of the Noordwijk Declaration and the follow-up is that carbon dioxide cuts should be around 20 per cent cuts on level of emissions in the year *2000* (see Figure 5.1), whereas Nordhaus is referring to reductions of some 6 per cent of carbon dioxide levels that would otherwise exist in *2050*. The implication is that the 'optimal' level of warming is significantly higher (and hence, emission reduction requirements are substantially lower) than is countenanced in the current international discussions, which we characterised as a constant capital approach. Cost-benefit analysis would be telling us not to take such drastic action.

There are some reasons for supposing that Nordhaus's analysis understates the degree of control required. First, his damage costs could be conservative if the potential for catastrophic events is realised. Table 5.3 demonstrates that if the damages are higher (the benefits of emission reduction are the same as avoided damages), then the 'optimal' level of control will also be higher. This is the essence of the 'constant capital' approach, which puts the emphasis on the degree of uncertainty and hence the potential for irreversible losses. How far the potential for catastrophic damage is borne out by the scientific evidence is, however, open to question. Second, the control cost estimates used by Nordhaus appear to ignore energy conservation measures that are cost-effective now without the need to introduce measures to raise energy prices through, for example, carbon or gasoline taxes. If this is correct, then Table 5.3 shows that lower reduction costs will mean a higher level of optimal control.[15]

What can we conclude about the cost benefit approach to global warming? At the very least, we can say that it should be pursued vigorously as a check on the economic implications of targets like those presented in Toronto. If the two approaches result in very different outcomes, then attention needs to be focused on why this is so. There is another reason for pursuing the cost-benefit approach. The Toronto target and the Noordwijk Declaration focus on *emissions reduction* as the appropriate response to global warming, but it is very unlikely that it will be efficient for policy to be formulated solely in terms of emissions reduction. Adaptation to global warming, through economic adjustment and defences against sea level rise, will surely play a part. The 'ecological limits' approach tells us nothing about the optimal combination of prevention and adaptation. Cost-benefit analysis is expressly formulated for just such a purpose. What a more detailed analysis would show, however, remains open to question.

ACHIEVING GLOBAL WARMING TARGETS

The Minimum Cost Principle

Whatever the target that is eventually set for an acceptable level of global warming, the remaining major issue is how it is to be achieved. An obvious point is that the targets must be internationally agreed. The United Nations Environment Programme (UNEP) is at work on a framework convention. This would be a statement which would recognise the problem and would signal an intention to take action. The setting of quantified targets would be a matter for one or more protocols. These protocols should be concerned with aggregate emissions of greenhouse gases rather than with carbon dioxide alone. The rationale here is that individual countries should be free to adjust their 'mix' of emissions to meet the target. This would keep down the costs of complying with the target. For example, if it is cheaper to cut CFCs per unit of radiative effect, then that is the first step that should be taken. If energy conservation is cheapest, then that should be pursued early on.

The idea of *minimising the costs of compliance* is important. This principle directs negotiators to find the most cost-efficient way of achieving a given environmental target. It is easy to misunderstand the principle, because it could be confused with getting environmental quality 'on the cheap'. This is not so. The idea is not to sacrifice environmental quality, but to achieve a quality target at the lowest possible cost. This releases resources for other purposes, including other environmental purposes. The minimum cost principle is also important when considering the probable shape of environmental policy over the next two decades. It seems very likely that all nations will face a rising bill for securing environmental quality. This is very much a legacy of our failure to take action during the past two decades – a reminder that precaution is better than reaction. But if the bill is going to rise anyway, then it is essential to seek the most effective means of minimising the increased costs. In this way, the cost burden to be borne by industrialists and consumers can be contained. This is not just a matter of wise husbandry, it is of strategic importance for the environmentalist case. Inefficient expenditures will risk alienating industry and, for that matter, consumers. By keeping costs down, the risks of a 'polluter backlash' can be minimised.

ALTERNATIVE ROADS TO CONTROL

In order to achieve internationally agreed global environmental targets, there are only three possible routes:

(1) 'Command-and-control' – pollution standards are set, and polluters are simply required to achieve that standard;

(2) Pollution taxes – polluters are taxed according to the level of emissions they produce;

(3) Tradeable permits – polluters are given permits to pollute up to the standard, but with the option of buying and selling permits in the market place.

The last two forms of regulation can be classified as *market-based incentives*. Their particular feature is that they make use of market signals. Pollution taxes involve: altering the prices faced by industry and, through these increases in industry's costs, the prices faced by consumers as well; or taxing consumers directly on the consumption of a polluting commodity. Tradeable permits operate through quantities, by allowing polluters to switch between sources of pollution provided they honour an overall target quality of the environment.[16] Command-and-control approaches, on the other hand, do not make use of the market at all.

Command-and-control

The three forms of regulation are not exclusive. In general, a sensibly designed environmental policy will be a mix of all three. Currently, and with odd exceptions, environmental policy is based on command-and-control. Regulation involves setting targets for those responsible for the initial emission of pollutants; this is the 'command'. The 'control' involves monitoring, inspection and penalty. Countries vary according to the intensity with which they pursue control. Self-evidently, command-and-control will not work if the penalties for non-compliance are small and less than the net gain from exceeding the standard. Typically, polluters prefer regulation: regulation is certain; polluters know where they stand; the standards are usually in place for some period of time; and, depending on the degree of control, polluters can usually negotiate with the regulator over special difficulties. There is something to be said for a system that reduces the uncertainty associated with regulation. But regulation is also more expensive than the other means of control and therefore offends the principle of minimum cost. There are two basic reasons why regulation is more expensive.

First, it requires that the regulator acquire information from the polluter, for example, about the costs of abatement technology. Second, it leaves the polluter with no flexibility. The polluter has to abide by the standard regardless of the fact that his or her costs of abating pollution could be much higher than those of another polluter. Because both polluters face the same standard, there is no possibility of the polluter with the lower costs of abatement taking a larger share of the control.[17]

We do have some idea of the excess costs of environmental regulation based on command-and-control. Tietenberg finds that the ratio of the costs of command-and-control approaches to market-based approaches in the US is high – that is, command-and-control is significantly more expensive.[18] Of 11 cases examined, four had ratios between one and two, five had ratios between two and 10 and two had ratios in excess of 10.

Compliance cost savings will be very important as the cost of protection rises. Expenditures on environmental protection in the Organization for Economic Cooperation and Development countries constitute between 1.5 and 2 per cent of their GDPs. To underline the prediction that these costs will rise, we may note that the Netherlands' National Environment Protection Plan anticipates a rise in this share to 3, or even 4, per cent. The Netherlands is one country which has formulated a detailed strategy for a marked improvement in its environment. Interestingly, it is based mainly on command-and-control policies. The percentage of GDP spent on environmental protection would almost certainly have been lower had they adopted market-based incentives, such as taxes and tradeable permits.

Some studies have begun to suggest that command-and-control may impose significant burdens on wealth creation in its more traditional form. Jorgensen and Wilcoxen estimate that the rate of growth of US GNP between 1973 and 1985 was 0.19 percentage points less than it otherwise would have been without environmental control.[19] Thus, a growth rate of, for example, 2.5 per cent per annum would have been reduced to 2.3 per cent per annum because of pollution control measures based almost entirely on command-and-control measures. It is slightly odd to focus on the impacts on GNP, because it is now generally agreed that GNP is a defective measure of changes in human welfare. Moreover, the benefits of environmental policy tend to show up in gains in welfare not captured by GNP – for example, long term health effects, gains in psychological welfare and improvements in amenity and biodiversity. Thus, the idea that the loss of a fraction of the GNP growth rate is a net cost of environmental policy must be resisted. What environmental policy has done is to switch a small amount of measured GNP into generally unmeasured gains in human welfare. Nonetheless, for good and bad reasons, politicians and others worry about the impact on GNP. The point here is not so much the adequacy or otherwise of GNP as a measure of human welfare, but relates to the very high probability that the costs of environmental regulation in the US could have been much lower had more use been made of market-based approaches.

Taxation Solutions

The principles underlying a pollution tax are readily understood. If the aim is to control, for example, carbon dioxide emissions, then emitters of carbon dioxide would be charged according to the carbon content of the fuels they burn. Coal would attract a higher charge than oil, which would attract a higher charge than natural gas. The effect of the tax would be to induce three things: substitution of lower carbon fuels for high carbon fuels, substitution of non-carbon energy for carbon energy (nuclear power and renewables) and energy conservation. The essence of the tax is to encourage tax avoiding behaviour. It is thus an incentive tax rather than a revenue raising tax. Nonetheless, revenues will be raised, and these can be used in a 'fiscally neutral' way to reduce income or other taxes. This softens any burden that may be believed to exist under the tax option, and it serves to reduce tax disincentives in the economy. Command-and-control procedures generate no revenue, so this option does not arise.

There are other virtues of a tax solution. For instance, because a pollution tax is a tax on all emissions, there is a systematic, continuing incentive to search for new technologies that will reduce the tax burden by removing the basis on which the tax is assessed – that is, pollution. Again, the command-and-control approach tends to lack any 'technology-forcing' characteristic, because it is invariably based on some concept of 'best available technology', although, in principle, standards can be set so as to be technology-forcing.

Tradeable Permits

As a final piece of evidence on the cost minimisation properties of market-based instruments, consider the limited experience there has been with tradeable emission permits under the Clean Air Acts in the US. A brief digression on the nature of tradeable permits is necessary, because their functioning is perhaps less straightforward than emission taxes.

Two broad sources of inefficiency arise in the command-and-control approach. First, command-and-control requires the regulator to use up resources to acquire information that polluters already possess. For example, polluters know far better than governments what it will cost to abate or clean up waste emissions. Yet, under the command-and-control approach, governments must obtain this information. Second, polluters vary in the ease with which they can abate pollution; stated another way, their costs of control differ. Under the command-and-control system, each polluter has to achieve a given standard, subject usually to some consideration about 'excessive' cost. Control is not concentrated in the sources that find it cheapest to abate pollution. Yet such a process of concentration would enable overall costs of compliance with the standard to be minimised.

The basic idea underlying tradeable permits is straightforward. First, an acceptable level of pollution is determined. This may be expressed as: some allowable concentration of, for example, lead in gasoline; a production or consumption target for chemicals, for example, CFCs; or an allowable national emission level, which is likely with carbon dioxide at some time in the future. Permits are then issued for the level of emissions up to the allowable amount. If, for instance, 100 units of pollution is allowable, then 100 permits (each with a value of one unit of emission) might be issued. There are various ways of determining the initial issue of the permits. Because of the disruption that might ensue by alternative allocations, a popular initial allocation is one based on historical emission levels. This is known as *grandfathering*: rights to pollute are based on past emission levels. However, this is not the only way to determine the initial allocation.

Once the initial allocation is made, polluters are then free to trade the pollution rights. It is this *tradeability* which is the hallmark of the permit system, because it is tradeability which helps to keep down the costs of complying with regulations. Tradeability is thus the main attraction of the permit system. Basically, a firm that finds it comparatively easy to abate pollution will find it profitable to sell its permits to a firm that finds it expensive to abate pollution. Essentially, it will sell the permit if it receives a price higher than the costs it will have to bear of abating pollution now that it has no permit. The high cost polluter, on the other hand, will find it profitable to buy permits if the price is below what it will otherwise cost him or her to abate pollution. Both low and high cost polluter, therefore, stand to gain, and this provides the incentive for them to trade. Moreover, by trading, the control of pollution will tend to be concentrated among those polluters who find it cheap to pollute, and permit holding will tend to be concentrated among those who find it expensive to control pollution. Yet the overall environmental standard is safeguarded, because nothing has happened to alter the overall number of permits, and it is this that determines the level of pollution.

Clearly, such a description is simplistic, but it captures the essence of the tradeable permit system. One important point to note is that trade need not be between *different* polluters. It can be between different sources within a single firm. The result is the same, however, because the firm will gain by concentrating abatement in its low cost sources and concentrating permits in its high cost sources.

If the above description is broadly accurate, then we would expect the actual experience of permit trading to result in no decline in environmental standards and a reduction in the costs of compliance compared to what would have been incurred in a command-and-control system. By and large, this is the experience in the United States, where a tradeable permits system exists as part of the US Clean Air Act.

Some terminology is needed in order to understand the US system. *Netting*, introduced in 1974, is a procedure whereby a firm can create a new emissions source provided it offsets the resulting emissions by reductions elsewhere in the same plant. Netting always involves *internal trading*, because the firm is not allowed to acquire permits from outside. *Offsets* were introduced in 1976 in areas where the Clean Air Act standards had not been met ('non-attainment areas'). Stringent rules applied to new sources would have meant that such areas could attract little or no new industry. However, by offsetting the new source by even greater reductions in existing sources, these areas are allowed to acquire new industry. Such offsets can be obtained by internal *and* external trading – that is, by buying up permits from within the same source or from other firms. *Bubbles* are perhaps the most famous part of the US tradeable permits system – they were introduced in 1979. A 'bubble' is a hypothetical aggregate limit for existing sources of pollution (whereas netting and offsets relate to new sources). Within the overall bubble limit, firms are free to vary sources of pollution so long as the overall limit is not breached. Bubbles are allowed to extend beyond a single firm, but in practice bubbles have tended to be placed round single firms. *Banking* was introduced in 1979 and operates just like a bubble but through time: the firm is allowed to bank credits and use them at some stage in the future. The US experience with bubbles, offsets, netting and banking is summarised in Table 5.4.

Table 5.4 raises a number of interesting points:

(1) Nearly all trading has been internal; only the offset system has resulted in moderate external trading. While the bubble system does permit external trading, hardly any has occurred. This appears to be due to the high costs of acquiring information about other firms' willingness to trade and the costs of obtaining the regulator's permission to trade.

(2) The cost savings are considerable, with a minimum of US$1 billion and perhaps as much as US$13 billion having been saved. A 'best guess'

TABLE 5.4 *US Experience with Tradeable Permits*

	Bubbles		*Offsets*	*Netting*	*Banking*
	Federal	*State*			
Number of trades	42	89	2 000	5 000–12 000	< 120
Cost savings US$ million	300	135	large	525–12 300	small
Air quality impact	zero	zero	zero	probably insignificant	probably insignificant
Nature of trade: internal	40	89	1 800	5 000–12 000	< 100
external	2	0	200	0	< 20

SOURCE: R. Hahn and G. Hester, 'Where Did All the Markets Go? An Analysis of EPA's Emissions Trading Program', *Yale Journal of Regulation* (Vol.6, No. 1, Winter 1989) pp. 109–53. See also R. Hahn, 'The Market for Bads: EPA's Experience with Emissions Trading', *Regulation* (Nos 3/4, 1987).

is US$10 billion.

(3) Banking has hardly been used at all.

(4) The extensive use of netting compared to bubbles, even allowing for the predominance of internal trades, is surprising, since bubbles apply to existing sources whereas netting applies only to new or modified sources.

Explaining the less than hoped-for level of trading activity in the US permits system is not easy. Commentators have suggested five main reasons:

(1) New sources are subject to far stricter regulations about emissions quality. This means that firms are keen to adopt any offsetting procedure when a new source starts up. Netting is the appropriate procedure in these cases, and this does much to explain the dominance of netting in the US system. Moreover, existing sources have inherited abatement equipment – bought before the bubble policy was introduced in 1979 – and therefore the costs of adjustment under a bubble policy are high.

(2) There is uncertainty about pollution credits. There has been considerable questioning about just what emission credits ensue under the banking legislation. Firms are not always sure how the regulator will determine baseline emissions and hence how emission credits will be determined. This uncertainty is heightened when other firms' credits are the subject of the trade – that is, when external trading is involved. Firm A has to be sure that firm B really will reduce emissions to create credits that can be traded. In contrast, internal trading involves far greater certainty, because the firm is dealing only with itself.

(3) It is more expensive to acquire information about external trading, because firm A needs to find out what other firms have banked credits and the price at which trade is likely to take place. Similar problems have arisen in other countries where attempts have been made to establish waste exchange information services.

(4) Firms will not trade with other firms if permit prices are expected to rise. Permits will be hoarded as long as the expected price rise is greater than the cost of hoarding – that is, greater than the interest rate.

(5) Hoarded permits can be used as a deterrent to new entrants.

In Europe, environmentalists have not been vocal in their objections to marketable permits. In large part, this is because they have had no specific applications to refer to. In the US, objections have been focused on two main issues: whether environmental quality is sacrificed under a tradeable permits system; and, whether it is morally right to 'permit' pollution, even for a price. Table 5.4 shows that the environmental quality argument has little or no foundation. There is no evidence that US environmental standards have been sacrificed because of the trading system. Of course, it could be argued that without the permits system standards might have had to be made tougher still, but this is an untestable proposition. Nor would it follow that tougher standards would be cost-effective, if it were true. The second objection has to be countered by an educative process. All regulatory systems 'permit' pollution if by pollution is meant waste. No economic process is waste free. Nor could it be, by the laws of thermodynamics. The issue has, therefore, to be one of whether a tradeable permits system somehow allows more waste than a command-and-control system. As we have seen, there is no reason at all for this to be the case. Objections to a tradeable permits system on the grounds of 'permitting' pollution are deficient, but the political reality is that such false arguments have influence. It is important, therefore, to secure well-informed debate on permit systems, as with other market-based incentives.

The US experience suggests that certainty about the regulatory system is highly valued. With command-and-control systems the firm is, by and large, clear about the nature of the regulation and what is and what is not permitted. This is also true for command-and-control systems that are less rigid, such as the UK system in which there is considerable scope for flexible adjustments in light of dialogue with the Inspectorate of Pollution. As European Community Directives play an increasingly important role in Europe, however, we might expect more and more 'standard setting' to replace the system of negotiation over achieving standards. It is clear, however, that in terms of the design of a tradeable permits system, it is important to minimise the kinds of uncertainty that appear to be endemic to the US approach.

Regulators will naturally be sensitive to the concerns of both environmentalists and industry. Nonetheless, they will also have their own concerns, primarily arising from the costs of considering, formulating and implementing any departure from the established command-and-control approach. It is worth remembering that the command-and-control mode of thinking is ingrained in environmental regulation in Europe, reflecting as it does the experience of over 100 years of public health, workplace and environmental legislation. Anxiety also tends to increase as less is known about the new system.

MARKET-BASED INSTRUMENTS AND GLOBAL ENVIRONMENTAL CHANGE

What is the relevance of market-based instruments to global environmental issues? A protocol on greenhouse gases will set a target for greenhouse gas reduction; but, just as it is inefficient to set each polluter the same target reduction in emissions, so it is inefficient to set each country the same target. There is a real danger that negotiators of a global warming protocol will make this mistake. For example, it seems fair for each country to achieve the stable emissions target of the Toronto Conference or a laxer one of stabilising emissions at current levels of carbon dioxide; but the fairness is illusory, because the target is set regardless of the different costs of achieving it. The aim should be to bias the reductions towards those countries that can most easily achieve them. The logic of this requirement is fairly simple: if one country has lower costs of abatement than another, then it will be cheaper to require more control in that country than in the high abatement cost country. A protocol that requires equal emission reductions by country offends this principle and hence incurs an unnecessary aggregate cost burden. It is conceivable that a protocol based on regulation could take account of the minimum cost objective, but it would be complex and, as it happens, unnecessary, because use of the tax or tradeable permits solution avoids the problem. Additionally, the tax and permit solutions help to solve the other dominant problem in international agreements, namely, how to devise incentives for co-operation.

Global warming affects different countries in different ways. A few states may conceivably gain from climate change; but, even if all lose, some will lose far more than others. Under these circumstances, it is going to be difficult to secure agreement on appropriate targets and on the initial allocation of emission reduction targets among countries. More importantly, the avoidance or containment of global warming is an example of what economists call a 'public good'. If global warming is reduced, then it will generally be of benefit to all countries, and no country can be excluded from the benefit. Any one country could secure the benefit of a global agreement without sharing in the cost. The United Kingdom, for example, could refuse to co-operate and wait for the rest of the world to solve the problem. After all, the UK contributes only 3 per cent of the world's emissions of greenhouse gases. By not co-operating, the UK could avoid the costs and reap the benefits and thus would be a 'free rider'.

The potential existence of free riders means that any protocol must have inbuilt incentives to encourage co-operation. We have already noted the problems of securing global co-operation on CFCs, and this is in the context of a comparatively straightforward agreement, because relatively few countries are involved. Barrett has drawn attention to the dissimilarity of the ozone layer case and global warming case, contrary to the widespread view that all we need to do for global warming is follow the example set by the Montreal Protocol.[20] Essentially, the Montreal Protocol is a game being played among very few players. Moreover, the costs of switching out of CFCs are comparatively modest. In order to persuade the potential free riders to co-operate, it is necessary to create incentives. This involves transferring resources – funds, technical assistance and technology – to these potential free riders. The scale of these transfers could be large, and it is far from clear that the world's leaders have understood the requirement to make transfers on the scale that is likely to be necessary.[21] But in the ozone layer case, the transfers are relatively modest. Markandya has estimated that perhaps US$1.8 billion to US$2 billion is needed to assist developing countries to develop substitutes for CFCs.[22] Very much larger sums will be needed to assist with substitution for coal, energy efficiency and so on. In short, the critical feature of a global warming protocol has to be the design of incentives for co-operation, and that means resource transfers.

It is the twin features of cost minimisation and resource transfer which make the use of market-based instruments attractive for implementing a global warming protocol. Both the tax and tradeable permits solutions need to be considered.

The International Tax Solution

In discussing an international tax solution we focus on carbon dioxide, but as noted earlier, it will be more efficient if any protocol allows individual greenhouse gases to be traded off in ratios determined by their radiative properties. Hoel has shown that a tax internationally administered and collected by some central agency is too bureaucratic and would interfere with domestic sovereignty. A tax implemented by each government would run foul of the free rider problem, because governments could easily offset a carbon tax by reducing other fuel taxes.[23] The solution, therefore, has to be one in which a central agency taxes each *country* according to emission levels. The tax level set would be the same for each country. Tax revenues would then be reimbursed; handed back to countries, but according to some formula of allocation. Each individual country would then act to minimise the sum of its tax payments and abatement costs. It would prefer to pay the tax if the cost of abatement is higher than the tax, but it would prefer to abate if that is cheaper than paying the tax. We would expect some combination of both actions.[24] The size of the tax would be determined by the agreed carbon dioxide emission reduction. Depending on the reimbursement rule chosen, some countries would then be net payers of tax to the central agency, while other countries would be net recipients of tax revenues. But the net payers of tax would still be better off under the agreement than they were without it, for they would have secured the benefits of the global warming damage that would have been avoided. Indeed, a condition of a successful agreement is that each country is better off with the agreement, taking into account net tax payments, abatement costs and environmental benefits.[25]

The rationale for reimbursement of tax revenues arises from the need to make resource transfers, as I have already demonstrated. Reimbursements should therefore relate to the costs of controlling carbon dioxide emissions and the damage likely to be experienced from global warming. The higher the cost of control, the higher will be the reimbursement required. Coal-based economies, for example, China and India, will tend to require large reimbursements. Countries that can switch easily from coal to gas-fired electricity production will have low reimbursement requirements. The United Kingdom might be such a country, but as noted earlier, the UK will have an incentive to focus a protocol on all greenhouse gases in order to minimise the extent that adjustments have to be made to carbon fuels and hence to the country's coal industry.

As far as damage is concerned, those countries that stand to lose most from warming, such as deltaic or island countries, are likely to co-operate because of their concern that the protocol should not fail. They will tend to require lower reimbursements from the tax revenues.

If, as in the ozone layer case, it is the developing countries which are most likely not to co-operate, then a formula that relates reimbursements to population would appear to be a prime candidate for the protocol. The resulting scale of transfers from rich to poor could be substantial and could arouse strong opposition, because some industrialised countries could be worse off after the tax than before it. Not only does Grubb doubt that an international fund with substantial carbon tax revenues would be acceptable,[26] but a team of New Zealand authors also questioned its political acceptability and efficiency.[27] However, Hoel thinks that such a fund is feasible and that reimbursements should be based mainly on GNP (and only partially on the basis of population) in order to secure co-operation.[28] Recent work in Canada suggests that a global carbon tax designed to secure the Toronto Conference ultimate target of 50 per cent reduction in carbon dioxide could result in tax revenues of US$600 billion, of which US$480 billion would be reimbursed to developing countries.[29] To gather some idea of the significance of these figures, US$600 billion is equal to 51 per cent of the entire external debt of the developing world in 1989 or one-and-a-half times the entire external debt of Latin America or six times the total of all disbursements to the developing countries in 1989. It is scarcely credible that any single international agency would have the capability to manage such resource transfers.

If a tax solution could be implemented, how big would the tax have to be? Barrett estimates that the long run tax (that is, the tax that would prevail once people were able to adjust by switching technologies and so on) would be approximately 54 per cent on the current price of coal, 43 per cent on the oil price (prior to the recent Gulf Crisis) and 32 per cent on the price of gas in the UK in order to achieve the Toronto aim of a 20 per cent reduction in the rate of 1988 emissions.[30] Short-run taxes (that is, not allowing for longer run adjustment) would be three times these levels. Another work by Ingham and Ulph suggests that any carbon tax would have to increase rapidly over time to substantial levels of over 100 per cent of average energy prices.[31] Whalley and Wigle suggest that a global tax rate of 90 per cent would secure 50 per cent reduction in carbon dioxide emissions.[32] There are also problems in setting the tax. It is well known that setting the initial tax may be a hit-and-miss affair. This may not matter much if there is scope for adjusting the tax at a later date. However, the problem with global warming is that delays in adjusting the tax will mean additional committed warming because of thermal inertia in the oceans. The adjustment process may therefore be quite complex.

An international carbon tax is not to be dismissed, but it is clear that, as with any solution, there will be formidable problems of design and implementation. The main reason for keeping such a tax on the agenda is that it is likely to be a more efficient instrument of control than country-by-country targets.

Internationally Tradeable Permits

The last option for implementing a protocol is through tradeable permits. The essence of the permit solution is that countries have an incentive to trade permits with each other and make net gains in the process. For example, suppose country A and country B receive the same level of quotas (that is, permits to release greenhouse gases), but that country A finds it much cheaper to control emissions than country B. Country A has an incentive to sell its quota to country B, collecting money from country B in excess of the costs of abating carbon dioxide, which it must now do, because it has no permits. Country B secures the permits and gains financially, because the cost of buying the permits is less than the cost it would otherwise have incurred by reducing carbon dioxide emissions. Both countries gain, and the overall emission target has not been compromised. In general, countries have an incentive to trade their permits until the marginal costs of abatement are just equal to the price of the permits in the market place. If costs exceed the price, then it will pay to try to buy further permits. If abatement costs are lower than the price of permits, then it will pay to sell the permits, collect the revenue from their sale and use some of the proceeds to abate emissions. Recalling the earlier discussion of tradeable permits, this is entirely analogous to the company trading of emission credits in the US, but with companies replaced by countries. A number of other trading systems exist: for example, fisheries quotas in New Zealand and Iceland and milk quotas in the UK.

Once again, some agency to oversee the trade in permits would be needed. Three major problems arise from this requirement. First, some countries are large emitters of carbon dioxide, notably, the US, USSR and members of the European Community. This may have some undesirable consequences for the efficiency of the tradeable permit approach, because the price of the permits on the open market will be influenced by sales and purchases by 'big' countries.[33]

Second, the tax solution penalises countries that emit too much, but the permits system requires additional penalties if countries go over their permitted level. This is a particularly relevant problem in the global warming context, because it is not clear what sanctions can be applied if countries persistently exceed their permitted emissions.

Third, the overriding problem involves the initial allocation of permits. This is akin to determining reimbursements under the tax solution. Few writers have favoured the grandfathering of permits, whereby they are initially allocated according to the existing levels of emissions. This would appear to favour the industrialised countries and does little or nothing to create incentives for the developing world to co-operate. But, in fact, any allocation other than grandfathering is likely to be resisted strongly by the major emitting nations. Without grandfathering, an agreement would be unlikely to come about. This suggests that, initially, a grandfathering approach would be most appropriate, but then emission allowances should be modified by altering the 'value' of the quota over time. Thus, a developed country quota would have declining allowances over time, while developing countries could have rising allowances that less than offset the developed nations' reductions. The burden of international agreement would be thus shifted from the initial allocation as emission allowances change over time. Countries are more likely to agree to this formula, because it allows time for adjustment and reduces the uncertainty attached to buying permits in the international market at the outset of the agreement.

Some authors have favoured a system whereby the initial allocation is based on GNP, but this would again favour the industrialised countries (as would GNP per capita). Per capita allocations have found most favour, and an initial allocation based on such a rule would have the effect of providing developing countries with a large quantity of permits which could then be sold to other countries at a profit over the cost of abating greenhouse gas emissions.[34] Grubb has suggested that in order to avoid the implicit reward for overpopulation, only adult populations should be counted. Another suggestion involves allocating all permits to non-polluters (that is, the world's poor) and then allowing trade thereafter. To avoid the hoarding of permits by rich countries, permits could be subject to renewal – that is, be subject to lease rather than ownership.[35] Countries with a capacity to create carbon sinks, such as new forests, could secure credits under the system – they might be allowed to emit beyond the level in their permits, provided the carbon sinks offset the excess.

The reality of international political economy is likely to work against any allocation of permits based on population. It would require assurance at the outset that international trade in permits would take place, otherwise the burden of adjustment would fall heavily and rapidly on the countries with high per capita emissions (for example, the United States). In these circumstances, such countries are unlikely to agree at the outset. Grandfathering of some kind is, therefore, likely to be the only initial allocation that will meet with agreement. The sheer 'newness' of tradeable permits on the international scene may, in any event, militate against them totally. If so, one essential message for international negotiators is that they should 'mimic' as best they can the efficiency of market-based approaches. To this end, allocated emission reductions must bear some resemblance to the pattern that would emerge if they were allocated according to costs of abatement. Even that requirement is a formidable one in terms of international negotiation.[36]

CONCLUSION

The problems of designing international agreements to meet the global environmental challenge are formidable. This paper has offered only the briefest of overviews of what is a major set of issues for immediate and future debate. The basic questions posed are highly problematic, namely, what is an acceptable level and rate of global warming and how can we design an efficient structure of global incentives for co-operation. The answers will involve more imagination and vigour than the international community has shown so far. It is evident that many of the existing discussions are based on the belief that natural science alone can determine what is acceptable. This is false. Social scientists also have an important part to play as global warming's impacts are measured and its costs are computed. However, assessing impacts is only one aspect of the decision-making problem, because the task of designing incentives is also pivotal. Humans are all too rational in their self-interest. It is this self-interest which will threaten international agreement, unless of course, we can use and manipulate self-interest to global advantage. This is the real global challenge.

Notes

This is a substantially revised version of The Henry Sidgwick Memorial Lecture which the author gave at the University of Cambridge, 23 February 1990. The author is indebted to Jonathan Fisher, David Fisk, Edward Barbier and Scott Barrett for comments on the earlier version. The author accepts responsibility for any remaining errors.

1. D. W. Pearce, A. Markandya and E. Barbier, *Blueprint for a Green Economy* (London: Earthscan Publications, 1989).
2. R. Solow, 'On the Intergenerational Allocation of Natural Resources', *Scandinavian Journal of Economics* (Vol. 88, No. 1, 1986) pp. 141–54; T. Page, *Conservation and Economic Efficiency* (Baltimore, MD: Johns Hopkins University Press, 1977); J. Hartwick, 'Intergenerational Equity and the Investing of Rents from Exhaustible Resources', *American Economic Review* (Vol. 66, 1977) pp. 972–4; and K-G Maler, 'Theoretical Foundations of the Concept of Sustainable Development', Seminar on *The Economics of Environmental Issues* (Paris: OECD, October 1989, unpublished).
3. See World Commission on Environment and Development, *Our Common Future* (Oxford: Oxford University Press, 1987).
4. See, for example, D. W. Pearce, E. Barbier and A. Markandya, *Sustainable Development: Economics and Environment in the Third World* (London: Edward Elgar, 1990).
5. One obvious issue is what measure of capital is being held constant: some 'physical' concept, real prices or real values.
6. See I. Mintzer, 'Cooling Down a Warming World: Chlorofluorocarbons, the Greenhouse Effect, and the Montreal Protocol', *International Environmental Affairs* (Vol. 1, No. 1, Winter 1989) pp. 12–25.
7. A very readable introduction to the global warming issue is S. Boyle

and J. Ardill, *The Greenhouse Effect* (London: Hodder & Stoughton, 1989).

8. Thus, speaking at the White House Conference on Science and Economics Research Related to Global Change, April 1990, Michael Boskin, Chairman of the President's Council of Economic Advisers, compared the cost of greenhouse gas emission reduction to the economic disruption caused by oil price rises in the 1970s and early 1980s. He remarked that 'available studies suggest that it would cost at least 1 per cent of annual US GNP, or perhaps several times that amount, to meet widely-discussed CO_2 reduction targets and that economic growth could slow significantly'. The conference itself was clearly aimed at persuading other countries to adopt more cautious attitudes, coming as it did before the publication of the reports of the UN Intergovernmental Panel on Climate Change (IPCC) and the Second World Climate Conference of November 1990.
9. *The Noordwijk Declaration on Climate Change* (Leidschendam, the Netherlands: Climate Conference Secretariat, 6–7 November 1989).
10. *The Changing Atmosphere: Implications for Global Security, Conference Statement* (Toronto, 27–30 June 1988). Paragraph 22 states (emphasis added):

    An *initial global* goal should be to reduce CO_2 emissions by approximately 20 percent of 1988 levels by the year 2005. Clearly, the *industrialized nations have a responsibility to lead the way*, both through their national energy policies and their bilateral and multilateral assistance agreements. About one-half of this reduction would be sought from energy efficiency and other conservation measures. The other half should be effected by modifications in supplies.

11. The limit of 0.1°C is suggested, for example, by International Project for Sustainable Energy Paths, *Energy Policy in the Greenhouse* (El Cerrito, CA: International Project for Sustainable Energy Paths, 1989).
12. In turn, the constant capital approach has similarities with the idea of adopting 'safe minimum standards' (SMSs). A SMS approach calls for a strong bias to conservation of natural capital unless the opportunity costs of the conservation – that is, the foregone benefits of developing the resource – are clearly very large. See R. Bishop, 'Endangered Species and Uncertainty: The Economics of a Safe Minimum Standard', *American Journal of Agricultural Economics* (February 1978) pp. 11–18.
13. See E. Barbier and D. W. Pearce, 'Thinking Economically About Climate Change', *Energy Policy* (January/February 1990) pp. 11–18.
14. W. Nordhaus, *To Slow or Not to Slow: The Economics of the Greenhouse Effect* (New Haven, CT: Yale University, Department

of Economics, mimeo., February 1990). For some modifications of Nordhaus's estimates, see J. Walter and R. Ayres, *Global Warming: Damages and Costs* (Laxenberg, Austria: International Institute for Applied Systems Analysis, mimeo., 1990); and R. Ayres and J. Walter, *Global Warming: Abatement Policies and Costs* (Laxenberg, Austria: International Institute for Applied Systems Analysis, mimeo., 1990).

15. Various studies suggest that there are very low costs of control through energy conservation. See Ayres and Walter, *op. cit.*, in note 14. For the United Kingdom, see Department of Energy, *An Evaluation of Energy Related Greenhouse Gas Emissions and Measures to Ameliorate Them*, Energy Paper No. 58 (London: HMSO, 1989). How far this observation affects the Nordhaus study is not clear, however, because Nordhaus captures energy saving measures by simulating the effects of a carbon fuel tax.
16. There is a formal relationship between taxes and the prices of tradeable permits. Both the tax and the permit are aimed at securing a predetermined level of pollution. Under conditions of certainty, it is then the case that the market price of a tradeable permit is equal (per unit of emission) to the tax.
17. The basic requirement for a minimum cost approach to regulation is that the different polluters' marginal costs of pollution abatement should be equalised. A tax that is the same for all polluters will achieve this result, because each polluter will prefer to abate than pay the tax if abatement costs are lower than the tax and will prefer to pay the tax if it is above abatement costs. A tradeable permit system has the same property. If marginal abatement costs differ, then there are gains to be obtained by having high abatement cost polluters buy permits from low abatement cost polluters.
18. See T. Tietenberg, 'Economic Instruments for Environmental Regulation', *Oxford Review of Economic Policy* (Vol. 6, No. 1, 1990).
19. D. W. Jorgensen and P. J. Wilcoxen, *Environmental Regulation and US Economic Growth* (Cambridge, MA: Harvard University, Department of Economics, mimeo., July 1989).
20. S. Barrett, *On The Nature and Significance of International Environmental Agreements*, (London: London Business School, mimeo., May 1989).
21. The transfer issue is emphasised in M. Grubb, *The Greenhouse Effect: Negotiating Targets* (London: Royal Institute of International Affairs, 1989).
22. See A. Markandya, *The Costs to Developing Countries of Joining the Montreal Protocol* (London: London Environmental Economics Centre, mimeo., 1990).
23. M. Hoel, *Efficient International Agreements for Reducing Emissions of* CO_2 (Oslo: University of Oslo, Department of Economics, mimeo., 1990).
24. The minimum cost theorem occurs again here, because the result will be

that each country will determine its combination of tax and abatement measures according to the rule that the tax rate, which is common to all countries, should equal the marginal cost of abatement. In this way, marginal costs are equated across countries, and this is the requirement for cost minimisation.

25. Unfortunately, it is more complex than this. The assumption here is that the global warming target is set in such a way that global environmental benefits exceed total abatement costs, in other words, that a cost-benefit rule is used. Only then is there a global 'surplus' to be redistributed. As we have seen, this may well not be the case. This raises the prospect that some countries will be net payers of tax and that their individual avoided damage costs will be less than the net payment. In these circumstances they are unlikely to co-operate. For this reason, it is important that some sort of cost-benefit rule is used for setting global targets.
26. Grubb, *op. cit.*, in note 21.
27. I. G. Bertram, C. Wallace and R. Stephens, *Economic Instruments and the Greenhouse Effect* (Wellington, New Zealand: Victoria University of Wellington, Economics Department, mimeo., August 1989).
28. Hoel, *op. cit.*, in note 23.
29. J. Whalley and R. Wigle, *Cutting CO_2 Emissions: The Effects of Alternative Policy Approaches* (London, Ontario: University of Western Ontario, Department of Economics and Waterloo, Ontario: Wilfred Laurier University, Department of Economics, mimeo., December 1989).
30. S. Barrett, London Business School, personal communication.
31. A. Ingham and A. Ulph, *Carbon Taxes and the UK Manufacturing Sector* (Southampton: University of Southampton, Department of Economics, mimeo., 1990).
32. Whalley and Wigle, *op. cit.*, in note 29.
33. See Hoel, *op. cit.*, in note 23.
34. Bertram *et al., op. cit.*, in note 27; and Grubb, *op. cit.*, in note 21.
35. Grubb, *op. cit.*, in note 21.
36. For further discussion, see D. W. Pearce, *Greenhouse Gas Agreements: Part 1 – Internationally Tradeable Greenhouse Gas Permits* (London: London Environmental Economics Centre, mimeo., 1990).

6 The Environmental Challenges in Eastern Europe

Craig ZumBrunnen

INTRODUCTION

Alvin Tofler's discussion of accelerated rates of change in his prophetic 1970 book, *Future Shock*, now seems almost pedestrian when compared to the actual pace, scale and pervasiveness of the revolutionary political, social, psychological, religious, artistic, moral, economic and environmental changes and events currently sweeping Eastern Europe and the Soviet Union.[1] Few predicted the relative ease and lightning speed with which nearly 70 years of Soviet/Stalinist authority could be and has been eroded. Only slightly more than a mere seven years after US President Reagan labelled the Soviet Union 'the evil empire', the world witnessed, on 3 October 1990, the final dramatic and symbolic end of the Cold War, namely, the formal reunification of East and West Germany. As the political and military clouds associated with East-West Cold War hostilities have rapidly dissipated from the European (and global) landscape, they have revealed a seriously disturbed, polluted and endangered environment throughout most of Eastern Europe, including the USSR. On the tragic side, Chernobyl has become the global symbol of this human/technological induced environmental disruption. On the positive side, true *'glasnost'* began only after and, in no small measure, as a direct result of this acute technological accident, which had immediate and could possibly have long-term international, ecological ramifications.

Because the actions of all countries contribute to (and all nations will undoubtedly feel the effects of) global environmental change, this issue concerns all states in the world. There is little doubt, therefore, that the countries of Eastern Europe are intricately linked into this interdependent relationship. With such large production of industrial pollutants, Eastern Europe contributes significantly to the build-up of trace gases in the atmosphere, and given the projected natural and social impacts, these countries will not be able to escape all of the deleterious effects. Thus it is clear that the states of Eastern Europe will have to play important roles in the international scientific and political discussions on global environmental issues. At this point in time, however, it is somewhat inappropriate to concentrate environmental concerns in the region upon the global issues. Rather, the area is now (and will continue to be in the immediate future) looking inward; states of the region are trying to focus their environmental efforts upon 'putting their own backyards in order'. No doubt, such action will have important ramifications for these larger issues, which may, in the future, lead these states into greater activity at the global level. The immediate concern, however, of the region, and therefore also of this chapter, is upon national and regional environmental issues in the countries of Eastern Europe.

Three major tasks are addressed in this chapter. First, the general dimensions and typology of the environmental problems in Eastern Europe will be examined. Neither space nor readily available spatially disaggregated data allow for a detailed discussion of all of the environmental problems in Eastern Europe, even excluding the USSR. Accordingly, this assessment will necessarily be crude, beginning with a modest attempt to compare some available environmental data for the United States with those of the countries of Eastern Europe. The second, narrower task will be to identify the environmental issues and concerns that have received the most media attention in these countries over the past few years. The third and final endeavour will be to suggest some of the environmental implications of the demise of the administrative-command economic systems in Eastern Europe. This task is more speculative and multifaceted. On the one hand, it is true that decades of Stalinist centrally controlled and administered-command economies in Eastern Europe led to inefficient and irrational resource management practices, which are, in large measure, responsible for most of the region's current environmental degradation problems. On the other hand, however, it seems that replacing these bureaucratic institutions with market institutions will not alone solve this region's environmental problems. In fact, some problems may even be exacerbated. From a broader global perspective, one needs to ponder whether divisive Cold War tensions will give way to more co-operative regional and international efforts towards environmental restoration and protection. Or, alternatively, will these past political-military tensions merely be replaced by equally divisive transnational pollution quarrels and intranational, interregional and broader international economic discord, in which environmental problems continue to receive relatively low priority?

TABLE 6.1 *Atmospheric Releases of Harmful Substances from Stationary Sources for the USSR and US in 1986*

	Dust (suspended particles)	*Sulphur Dioxide*	*Carbon Monoxide*	*Nitrogen Oxides*	*Lead*
Total in USSR (in millions of tonnes)	16.5	18.8	15.4	4.3	0.01
Total US (in millions of tonnes)	5.4	20.3	18.4	10.8	0.0078
USSR (in tonnes per square km)	0.74	0.84	0.69	0.19	0.00045
US (in tonnes per square km)	0.57	2.13	1.93	1.13	0.00082
Ratio USSR/US (total)	3.06	0.93	0.84	0.40	1.28
Ratio USSR/US (per square km)	1.30	0.39	0.36	0.17	0.55

SOURCE: Calculated from data in *Okhrana okruzhayushchey sredy i ratsional'noye ispol' zovaniye prirodnykh resursov v SSSR* (Moskva: Goskomstat SSSR, 1989) p. 84.

CURSORY RETROSPECTIVE

Scholarly discussions of environmental problems in the USSR are not new and are rather extensive.[2] The environmental problems in Eastern Europe, however, appear to have received less attention until recently. Ivan Volgyes's 1974 edited volume, *Environmental Deterioration in the Soviet Union and Eastern Europe*, appears to be one of the earliest books to focus on environmental problems in both the USSR and Eastern Europe.[3] In his contribution to this volume, Leslie Dienes presents a concise, but comprehensive, geographic overview of environmental disruption in Poland, Czechoslovakia and his native Hungary.[4] The environmental topics he addresses include the air, water, land, soil and vegetation resources of these three East European countries, and he also analyses the influences of economic and spatial structures. Institutional factors, such as the role of state ownership of resources on questions of environmental resource management, are considered. He concludes with an East European perspective on the key global debate of the early 1970s, namely, demographic and economic growth versus the environment. Undoubtedly, Joan DeBardeleben's 1985 tome, *The Environment and Marxism-Leninism: The Soviet and East German Experience*,[5] and Barbara Jancar's *Environmental Management in the Soviet Union and Yugoslavia*[6] still represent the most detailed and comprehensive comparative analyses of the environmental debates in the Soviet Union and any of its (former) East European allies. Furthermore, the late George Hoffman draws upon a number of recent scholarly writings to compose a pithy overview of environmental problems in Eastern Europe, including Yugoslavia.[7] His follow-on discussion of the resource dependence and energy situation in Eastern Europe is also highly relevant to understanding the environmental pollution problems in this region, especially air pollution and acid precipitation associated with the high use of indigenous, low-quality, brown coals.[8] Examples from some of the above works will subsequently be referred to in the general discussion of the environmental challenges in Eastern Europe.

REGIONAL AND INTERNATIONAL ENVIRONMENTAL COMPARISONS: A FIRST APPROXIMATION

As previously indicated, a comprehensive examination of environmental conditions in Eastern Europe, excluding the European USSR, goes well beyond the stated tasks of this exploratory work. Nonetheless, it is critical to have some, at least relative or comparative, environmental 'benchmarks' at the outset. Recently published Soviet and Council for Mutual Economic Assistance (CMEA) environmental quality and resource use data allow for this cursory comparative appraisal. Tables 6.1 and 6.2 contain data, though highly spatially aggregated, which are useful for international comparative purposes.

Crude recent comparisons between the Soviet Union and the United States with regard to the atmospheric releases from stationary sources of five key harmful substances are listed in Table 6.1. The first two rows represent the total 1986 airborne releases in millions of tonnes for the USSR and US, respectively. More meaningful are the third and fourth rows of data, which represent these atmospheric releases in terms of average tonnages per square kilometre for each country. The last two rows, however, are most useful for comparative purposes. For example, in 1986, the Soviet Union's economy generated and discharged slightly

TABLE 6.2 *Atmospheric Releases of Harmful Substances from Auto Transport in the US and Selected CMEA Countries in 1986* *

US	*USSR*	*Hungary*	*Poland*	*Czechoslovakia*
Totals in millions of tonnes:				
48.9	36.2	1.23	1.58	0.53
Expressed in terms of tonnes/square kilometre of area:				
5.14	1.63	13.22	5.07	4.15

*Based on CMEA data for the sum of carbon monoxide, carbon dioxide, nitrogen oxides, hydrocarbons and, for Poland, also including sulphur dioxide and lead.

SOURCE: Calculated from data in *Okhrana okruzhayushchey sredy i ratsional' noye ispol' zovaniye prirodnykh resursov v SSSR* (Moskva: Goskomstat SSSR, 1989) p. 85.

more than three times the suspended solids, about 1.3 times as much lead and nearly as much sulphur dioxide and carbon monoxide as the United States, but with an economy producing perhaps only one-half to two-thirds as large a GNP. When these aggregate ratio data are expressed in terms of average concentrations per unit area (last row of Table 6.1), the data appear to be more reflective of the relative sizes of the two economies. The vast area of the USSR compared to that of the US, however, accounts for this apparent dilution of total Soviet industrial air pollution discharges. If one were to exclude both the vast stretches of the USSR in the southern deserts, the north and Siberia, which are sparsely inhabited, and the similar regions in the United States, then the comparison would be much starker indeed.

National aggregate air-borne discharges from non-stationary or auto-transport sources for the US, USSR, Hungary, Poland and Czechoslovakia are listed in Table 6.2. Two items should be noted. First, despite having a very much smaller auto-transport fleet, the Soviet Union generates by mass nearly 75 per cent as much air-borne pollutants as the United States. This reflects a number of factors, such as the larger fraction of the auto-transport fleet being trucks as opposed to automobiles, inefficient and poorly maintained engines, low-quality fuels and the lack of conversion to non-leaded gasolines in the Soviet Union. For comparative purposes, Table 6.2's derived average concentration data (row 2) have some value. The relative auto-transport emissions in Hungary exceed those of the United States by a factor of nearly 2.6. Poland and Czechoslovakia, both with proportionally much smaller auto fleets than the US, have comparable average emissions per unit area. Again, the vast area of the Soviet Union masks the real air contamination levels, especially in its larger urban areas, where both industries and trucks are concentrated.

The official CMEA data on harmful air pollution discharges for selected individual cities in Poland, Czechoslovakia and the European USSR, listed in Table 6.3, are difficult to interpret in a straightforward manner. If accurate, however, they reveal a somewhat surprising difference in the share by pollutant source between the USSR and the other two CMEA countries. According to these data, stationary sources of harmful atmospheric pollution dominate in Poland and Czechoslovakia, whereas auto-transport sources account for much larger fractions in the Soviet cities. This disparity may be explained primarily in terms of the structural differences in the production and consumption of energy and fuels between these two CMEA countries and the USSR, as evidenced in Table 6.4. First, Poland and Czechoslovakia are much more dependent on the burning of indigenous solid fuels for their industrial and space-heating needs than the USSR, which consumes much more oil and natural gas. Natural gas usage for space heating has grown rather rapidly in the urban areas of the European USSR in the past two decades.

Furthermore, the structure of energy production and consumption in all Eastern European countries has a great deal of inertia. For example, the geography of the energy resource endowments in the region dictates that only Hungary and Romania have significant natural gas production – and only Romania and Albania in terms of oil production – in relationship to their own domestic consumption needs (see Table 6.4). As the third column of Table 6.4 discloses, solid fuels dominate the energy production picture in all of the region except for Albania, relatively oil- and gas-rich Romania and the Soviet Union. To a lesser extent the same is true for energy consumption in the region (Table 6.4, column 4). This situation, of course, is a major cause of much of the environmental deterioration of the landscapes of Eastern Europe. Except for Poland and the USSR, much of these solid fuels consist of low-quality, soft-brown coals that have high-ash contents. Despite the large regional energy deficits in Eastern Europe (see Table 6.4, columns 1 *vs.* 2, and last row), the conversion in the near-term to cleaner oil and natural gas appears increasingly problematic for two reasons. First, as already noted, domestic supplies of liquid fuels and natural gas are very limited, even in Romania. Second, it should be noted that the economic crises and market reforms underway in the USSR and Eastern Europe indicate that the Soviet Union will intensify demands for increasingly scarce hard currency for its oil and gas exports to Eastern

TABLE 6.3 *Official CMEA Data on the Total Quantity of Harmful Substances Discharged into the Atmospheres of Selected Cities in CMEA Countries in 1987 (in thousands of tonnes)*

Country / *City*	*Discharge of Harmful Substances* *Total*	*Discharges by Type of Source* *Stationary*	*%*	*Auto-transport*	*%*
Poland					
Warsaw	235	187	79.6	48*	20.4
Krakow	618	596	96.4	22*	3.6
Wroclaw	93	64	68.8	29*	31.2
Gdansk	87	69	79.3	18*	20.7
Lodz	127	86	67.7	41*	32.3
Czechoslovakia					
Prague+	167	133	79.6	34	20.4
Bratislava	89	58	65.2	31	34.8
European USSR					
Moscow	1211	369	30.5	842	69.5
Leningrad	626	254	40.6	372	59.4
Minsk	235	110	46.8	125	53.2
Vilnius	97	37	38.1	60	61.9
Kiev	326	93	28.5	233	71.5
Kishinev	133	42	31.6	91	68.4

* calculated as a residual by author
+for centre of city

SOURCE: *Okhrana okruzhayushchey sredy i ratsional' noye ispol' zovaniye prirodnykh resursov v SSSR* (Moskva: Goskomstat SSSR, 1989) p. 85.

Europe. Furthermore, the Soviet Union may well find it impossible to continue surplus production of oil in particular.

Besides the highly visible polluted skies of most major East European coal-based industrial cities, one of the most extensive and observable environmental problems in the region is forest damage. An IIASA (International Institute for Applied Systems Analysis) study concludes that pollution damage alone to European forests amounts to at least US$30 billion annually. This loss is equivalent to the annual output of West Germany's iron and steel industry or, ironically, approximately three times the current total annual expenditures that Europe's political leaders have committed to controlling air pollution.[9] Much of the forest vegetation damage is the result of acid precipitation associated primarily with coal-burning power plants and industries in both Eastern and Western Europe. Because of the prevailing westerly to southwesterly winds over much of the continent, Eastern Europe is impacted not only as a result of its own high sulphur dioxide, coal-burning economies, but it also receives some of the acid precipitation originating from the coal-burning power plants and chemical factory smokestacks of Western Europe, especially from western Germany. Conifer species have been impacted most seriously. The symptoms include yellowing of needles, damage to fine rootlets and casting off of older needles. Biogeography and physical geography help to explain why conifers are hardest hit, namely, these needle-bearing species tend to be indigenous to high (or mountainous) elevations, where orographic lifting produces the highest precipitation levels.

As Table 6.5 reveals, however, the forests of Eastern Europe are not the only European forests experiencing serious human-induced environmental stress. In fact, the fraction of total forested areas damaged in Finland, Luxembourg, the Netherlands, Switzerland and West Germany all exceed the fractions for East European countries. On the other hand, some of the most severely damaged trees and forests are in Eastern Europe. For instance, areas of dead and dying trees amount to 450,000 hectares in Poland, or 20 per cent of the total forest area damaged in the country. In Yugoslavia, where overall 10 per cent of the forests exhibit damage symptoms, the percentage jumps to about 33 per cent in the relatively highly industrialised Republic of Slovenia.[10] Probably the most widespread serious forest wasteland is located in the Erzgebirge Mountains (or Krusne Hory Mountains), which straddle the former East German-Czechoslovakian border northwest of Prague. Near the dead and dying forest in the vicinity of the Czech industrial city of Most, where power plants burn high-sulphur coal, the sulphur dioxide concentrations have been reported to average 112 mg/cubic metre, or 13 times as high

TABLE 6.4 *Production and Consumption of Primary Energy in Eastern Europe in 1986*
(P) Production (C) Consumption

	Total		*Percentage of Total*							
	*Mtoe**		*Solids*		*Liquids*		*Natural Gas*		*Electricity*	
Country	*(P)*	*(C)*	*(P)*	*(C)*	*(P)*	*(C)*	*(P)*	*(C)*	*(P)*	*(C)*
Albania	4.5	2.8	18.1	34.9	67.3	43.7	8.0	12.9	6.6	8.5
Bulgaria	14.1	36.9	88.1	45.5	2.1	35.5	1.0	14.7	8.8	4.3
Czechoslovakia	47.1	68.4	94.5	65.4	0.3	18.5	1.2	13.2	4.0	2.9
East Germany	70.1	93.3	93.5	74.1	1.7	16.4	3.3	8.2	1.6	1.3
Hungary	15.1	27.9	39.9	30.6	19.0	30.3	36.8	33.5	4.3	5.6
Poland	124.6	123.3	96.6	81.6	0.2	10.5	3.0	7.7	0.3	0.3
Romania	63.0	72.7	23.6	26.8	17.5	19.2	57.2	52.3	1.6	1.8
Yugoslavia	24.7	39.8	63.8	47.0	17.2	33.0	7.3	12.7	11.6	7.3
USSR	1 581.1	1 258.0	23.2	28.2	39.0	29.0	35.8	40.4	2.0	2.4
Eastern Europe (excluding Yugoslavia)	338.5	425.3			15.4	24.9	15.8	20.4	3.9	3.5

* Mtoe = millions of tonnes of oil equivalent

SOURCES: United Nations data as adapted from George W. Hoffman, 'Eastern Europe', Chapter 10, in George W. Hoffman (ed.), *Europe in the 1990's: A Geographic Analysis* (New York: John Wiley & Sons, 1989) p. 612.

TABLE 6.5 *Estimated Forest Damage in Selected Countries of Europe, August 1986*

Country	*Total Forest Area*	*Estimated Area Damaged**	*Portion of Total Area Damaged*
	(thousands of hectares)		*(per cent)*
Eastern Europe			
Czechoslovakia	4 600	1 200	26
East Germany	2 900	350	12
Hungary	1 670	184	11
Poland	8 677	2 273	26
Yugoslavia	9 500	1 000	10
Western Europe			
Austria	3 754	960	26
Belgium	616	111	18
Finland	19 400	6 790	35
France	15 075	280	+
Italy	6 363	318	5
Luxembourg	82	42	51
Netherlands	309	155	50
Norway	8 330	400	5
Sweden	26 500	1 000	4
Switzerland	1 200	432	36
West Germany	7 371	3 824	52
Total	136 034	19 319	14

* estimates made in 1984, 1985 or 1986.
\+ 24 per cent of the 1.16 million hectares surveyed exhibit damage.

SOURCE: *Allgemeine Forst Zeitschrift* (Munich, West Germany, No. 46, 1985 and No. 41, 1986).

as those in an apparently unaffected forest situated some 160 kilometres to the southeast. Furthermore, it is claimed that peak sulphur dioxide concentrations are even several times higher.[11] Sulfur deposition levels are so high in parts of Czechoslovakia – more than 30 times the damage threshold – that current European commitments to lower European sulphur dioxide emissions by the year 2000 by 28 per cent from 1980 levels would have little salutary effect. Furthermore, calculations made by IIASA's simulation model RAINS indicate that these current reduction commitments in sulphur dioxide emissions would only yield a net reduction of 18 per cent.[12]

As if this degradation were not bad and long-term enough, a Czech geochemist, Tomas Paces, has found perhaps even more serious and possibly irreversible deleterious soil changes. His detailed chemical analyses of runoff from the Erzgebirge Mountains suggest that acidification has profoundly altered the soil's ability even to sustain a forest. He claims that less than half of these elevated nutrient losses can be explained by high atmospheric precipitation rates and thus atmospheric inputs. Rather, the losses of the buffering agents (for example, calcium and magnesium) have been causing normally bound up aluminum to be released. In its soluble forms, aluminum can be toxic to trees. Finally, Paces believes that the higher nitrogen concentrations in the runoff reflects the damaged forest's inability to recycle nitrogen properly.[13]

All of these dismal lines of evidence point to the real and inescapable possibility of widespread negative and irreversible ecological damage to the forests of the region. Diseased and dying forests, especially in mountainous areas, can lead to increased levels of destructive soil erosion and significantly higher risks of floods and avalanches. In the early 1980s, the Czechoslovakian Academy of Sciences estimated that the annual costs of acid precipitation pollution were US$1.5 billion, much of this attributable to forest damage costs.[14]

A final sobering thought is implicit in Table 6.5, namely, the process now underway of transforming the economies of Eastern Europe and the Soviet Union from administrative-command systems to mixed-market ones will not alone solve these direct and indirect air pollution problems. Even if, by some miraculous set of circumstances, the East European countries could eliminate their domestic sources of acid pollution, geographic wind patterns would still generate significant transfrontier pollution fallout from West European industries. Thus ironically, any environmental benefits arising from factory and plant closures ensuing from the economic disruptions of the current socio-economic transformation processes could well be only short-term. In turn, significant acid pollution originating from the industries and power plant smokestacks of Eastern

Europe carries downwind to the Soviet Union and, to a lesser extent, Sweden and Finland. Supposedly 5 to 10 per cent of the pollution falling on Finland comes from Poland.[15] This combination of physical geography and industrial geography presents some interesting theoretical and economic-ecological propositions. On the positive side, Finland and Sweden both apparently believe that it will be economically efficient and ecologically sound for them to provide East European industries with financial and technical pollution abatement assistance, so that they might reduce their own acid pollution damage costs. In point of fact, in May 1990, Finland's Minister of the Environment, Kaj Barlund, and the Polish Minister of Environmental Protection and Natural Resources signed a bilateral agreement on co-operation in environmental protection. Interestingly enough, a key innovative provision of this agreement exempts Poland from paying 90 million Finnish marks (approximately US$28 million) of its owed debt over three years; in exchange, this sum must be earmarked for ecological investments in Polish industry. The agreement

TABLE 6.6 *Data on Water Withdrawals in Selected CMEA Member Countries in 1987*

	Bulgaria	*Hungary*	*Poland*	*Czecho-slovakia*	*USSR*
Total Water Withdrawals (millions of cubic metres)	7 433	7 450	15 456	5 369	339 490
Portion from surface water sources (millions of cubic metres)	5 040	5 590	13 062	3 937	305 261
Portion from surface water sources (in per cent)	67.8	75.0	84.5	73.3	89.9
Per Capita Water Withdrawals (cubic metres per year)	829	703	410	345	1 199

SOURCE: Derived from *Okhrana okruzhayushchey sredy i ratsional'noye ispol'zovaniye prirodnykh resursov v SSSR* (Moskva: Goskomstat SSSR, 1989) p. 63.

TABLE 6.7 *Data on Waste Water Discharge for Selected CMEA Member Countries in 1987*

	Bulgaria	*Hungary*	*Poland*	*Czecho-slovakia*	*USSR*
Total Discharge of Waste Water (millions of cubic metres)	1 502	6 000	12 754	3 403	158 993
Waste Water Discharge as a Percentage of Total Withdrawals	20.2	80.5	82.5	63.4	46.8
Polluted Portion of Discharged Waste Water (millions of cubic metres)	266	100	1 722	632	20 620
Polluted Portion as a Percentage of Total Discharge	17.7	1.7	13.5	18.6	13.0

SOURCE: Derived from *Okhrana okruzhayushchey sredy i ratsional' noye ispol' zovaniye prirodnykh resursov v SSSR* (Moskva: Goskomstat SSSR, 1989) p. 63.

moreover calls for the transfer of Finnish ecology-serving technologies.[16] Two weeks later, Poland and Sweden signed a similar environmental agreement in which Poland will receive 300 million Swedish kronor (approximately US$45 million) for air pollution abatement projects in Poland.[17] Unfortunately, it appears that Western Europe has by no means solved its own acid-making emission problems and hence, would have little incentive to aid Eastern Europe's or the Soviet Union's pollution control efforts. 'Downwind' Russia has its own problems, especially the extensive Chernobyl clean-up costs, and it is certainly in no position either economically or technically to come to the assistance of obsolete, 'upwind', coal-burning factories and power plants in Eastern Europe.

Water resource quality and quantity problems are serious throughout Eastern Europe. Examination and comparison between Tables 6.6 and 6.7 allow for some crude assessments. First, implicit in the third row of Table 6.6 is the relative importance of surface water quality to each of the five countries for which Goskomstat publishes data. Bulgaria's relatively low reliance on withdrawals from surface water supplies suggests that its municipal and rural infrastructure is backward. In other words, the country is likely to rely more on individual small underground and open shallow dug wells for water supplies than in the four other countries listed. The per capita water withdrawal figures (Table 6.6, row 4) cannot be compared easily. For instance, while the Soviet Union has the highest per capita water withdrawals, it also uses by far the highest amount of irrigation water. On the other hand, multiple reuse of the Danube's water by Bulgaria, Hungary (and Romania) may allow them, indeed, to abstract on a per capita basis essentially twice the volume of water of Poland, Czechoslovakia and the former East Germany. The most important implication is the high share of surface water use, and hence, the existence of a real need to protect surface water quality.

In terms of ranking water pollution problems in the region, Table 6.7 is useful. The second row data are derived by expressing the country by country data for total discharge of waste water in terms of their percentage of the total water withdrawal (Table 6.6, row 1) for each country, respectively. Clearly, there is a wide range in this percentage for the available CMEA – Goskomstat country-level data, varying from Bulgaria's low figure of 20.2 per cent to Poland's 82.5 per cent. What factors account for this wide variation? First, as previously indicated, the Soviet Union has the highest irrigation or 'consumptive' water use of the 'former' CMEA countries. Second, it seems quite plausible that Bulgaria has the least developed centralised sewerage network of the five countries catalogued. Accordingly, the figure of 20.2 per cent may greatly understate the actual total percentage returned, that is, if discharges from central sewer outfalls as well as effluents from other non-centralised point and non-point effluent sources are considered. Frankly, the percentage of the total discharge of waste water reported as polluted (see Table 6.7, row 4) is rather enigmatic. If accurate, Hungary's surface waters should be perhaps an order of magnitude cleaner than those of the other countries. From long experience with such data from the Soviet Union, this author believes that 'polluted' most likely means 'waste water discharged without any purification treatment', whereas 'waste water' most likely means 'discharged after undergoing only primary treatment' that would not be included in the 'polluted portion . . . ' category. Furthermore, the severity and types of contaminants do vary greatly. Even using the old adage 'the secret to pollution is dilution (by at least a factor of 10)' as a standard would leave the quality of all but Hungary's waters in quite questionable conditions. Multiple reuse by downstream users makes the real situation worse, notably for such intensively used river waters as the Elbe in the erstwhile East Germany.

CURRENT ENVIRONMENTAL CHALLENGES AND ISSUES IN THE COUNTRIES OF EASTERN EUROPE

Let us now turn from the attempts to make crude international comparisons of environmental problems to a cursory review of the printed media in the various countries of Eastern Europe to reveal the current environmental debates and problems as they are being discussed in the media. These brief environmental vignettes serve to complement the material and fill in some of the gaps of the previous section. Throughout the region, environmental deterioration has generated independent environmental movements; hence, ecological problems have challenged the legitimacy of central-administrative economic institutions and governments throughout the region. Our media environmental highlights begin with the recently 'disestablished' East Germany.

The Environmental Deterioration Challenge in Eastern Germany

In January 1988, unofficial ecological groups began uniting into a movement called 'Ark-Green Network'. By November 1988, approximately 150 such unofficial environmental groups had come under the umbrella of the 'Network'. During the summer of 1988, East German ecologists secretly made a film entitled 'Bitter Things From Bitterfeld'. The camera's lens found ecological disasters everywhere, including an East German dumping ground scheduled to receive 2 million tonnes of West German wastes from the end of 1989. Serious public outcries were heard on both sides of the common German border once knowledge of this 'waste export' plan became known.[18]

Although the Goskomstat volume used to create many of the previous tables does not contain air pollution data for East Germany, DeBardeleben cites West German sources that claim that the cities of Leipzig, Zwickau and Halle have dust and sulphur dioxide contamination levels double those commonly found in Dusseldorf, Tokyo or Chicago. Furthermore, in the industrial region around Halle, the incidence of respiratory and heart diseases is 10 to 15 per cent greater and life expectancy is reportedly five years less than elsewhere in East Germany.[19] Again, these air quality problems are directly linked to East Germany's burning of high-sulphur, brown coal.

The former East Germany is poorly endowed with fresh water resources, particularly when one compares the region's water supply with its industrial and municipal water demand. The Elbe's flow is reported to be reused more than two and one-half times before its tainted waters reach the sea.

DeBardeleben notes that the waters of the southern Halle and Dresden districts as well as the Werra River are also heavily polluted.[20]

Two other issues associated with the 'Green Movement' in East Germany have been prominent in the press. First, provoked, perhaps, by the Chernobyl nuclear accident, East German ecologists and environmentalists have strongly criticised East Germany's nuclear power programme and its safety record. In January 1990, the press in both East and West Germany began writing about a December 1975 accident at the Griefswald nuclear power plant near Rostock – a plant which uses a Soviet-supplied VVER-440 reactor. By the middle of 1990, significant East and West German co-operation in the area of nuclear safety was in evidence.[21]

The second interesting 'Green Movement' issue was the short-lived political alliance between the Independent Women's Union (UFV) and the Green Party in the 1990 spring elections. The combined UFV/Green Party only garnered 1.96 per cent of the vote – this gave them eight seats in the People's Chambers. However, the electoral alliance was broken by the UFV when the Green Party refused to allocate three of the eight seats to UFV candidates.[22] Thus, despite the Green Party's activism and media prominence, it appears that economic and political factors other than the environment influenced most voters' decisions.

Finally, economic reconstruction and environmental clean-up costs in the former East Germany are truly staggering. Reconstruction in the energy sector alone is given a ten-year figure of DM50 billion, excluding the billions needed for ameliorating environmental damage.[23] Over the next decade, perhaps hundreds of billions of deutschmarks will be needed.

The Environmental Deterioration Challenge in Poland

Poland, like its neighbours, is engaged in heated debates surrounding its possible energy futures. The past and present desolation of a coal-based industrial infrastructure are nearly ubiquitous in Poland. But one megaWatt nuclear power plant capacity is reported to cost 290 million zloties in 1986 prices, compared to 100 million zloties for a conventional power plant consuming hard coal (excluding the desulphurisation costs). Furthermore, it takes three times as long to build a nuclear power plant.[24] So, Poles grapple with the emotions, the fears and the unclear economics of both an 'if' safer and cleaner nuclear future, and a dirtier and cheaper coal route. Having been in the path of Chernobyl fallout, Poles have some appreciation of just how important that 'if' is, and they are demanding referenda on the construction of nuclear power plants.[25] In June 1990, alternative energy advocates demonstrated against the formation of a French-Belgian consortium to complete the 'Zarnowiec' nuclear plant.[26]

Poland is immersed in a number of air and water pollution problems with both Czechoslovakia and Germany. Poland is very concerned about current and future transfrontier air pollution from a number of Czech factories. The most important of these include the large coking plant being built at Stonava, scheduled to be completed in 1992. This new facility will add to the existing discharges from the Trinec metallurgical combine and the Detmarovice power station. Czech authorities have tried to calm the concerned Poles by claiming both that the opening of the Stonava coking plant will be shortly followed by the closing of the older and nearby Karvina and Ostrava plants and that natural gas will be transmitted to Karvina. The Odra River and its neighbouring agricultural lands are polluted by oil originating in Czechoslovakia – an especially severe pollution episode occurred in this region in 1986.[27] In April 1990, Poland and Czechoslovakia signed an environmental agreement aimed at reducing pollution of their border rivers. Financial compensation, however, for the accidental pollution of the Odra River in 1986 remains unsettled, as Czechoslovakia has balked at the Polish side's damage claim of over 100 million clearing rubles.[28]

Poland and East Germany have had long-standing co-operative agreements dealing with the protection of their border waters.[29] Controversy has flared-up in the Bogatynia region of Upper Silesia over an agreement, signed 16 August 1988 between the authorities of Jelenia Gora Voivoship and German officials, to import 900,000 tons of highly polluted Rhine River sludge into the Bogatynia area. Once they became aware of the agreement, the local residents and 'Greens' became incensed that their local environs were becoming a dumping ground for West German industrial wastes.[30] Antiquated industries burning high-ash and high-sulphur coals are the technological basis for the environmental deterioration crisis in Poland.

The Environmental Deterioration Challenge in Hungary

Although there are some nuclear power issues in Hungary, water resources have been receiving highest environmental priority in the country.[31] The 1986–1990 Hungarian Five Year Plan, presumably the last, contains a special section on 'Environmental Protection and Nature Conservation'. In addition to the important emphasis the Plan gives to the shrinking water resources along the Danube River in the Middle-Transdanubia area, the Sajo Valley and the environs around Pecs, protection of arable land is emphasised. Other serious problems, such as the processing and disposal of communal wastes, urban air pollution, the reduction of sulphur dioxide emissions and the protection of relatively unspoiled environments, are also mentioned.[32]

Seemingly by far the most notable environmental issue over the past several years in Hungary has been the proposed Gabcikovo-Nagymaros hydroelectric project. This controversial project is the result of a 1977 Hungarian-Czechoslovakian agreement to construct jointly a hydroelectric and navigation system at the two ends of the 120-kilometre-long Gabcikovo-Nagymaros section of the Danube, which forms a common border. Under a separate bilateral contract, Austria agreed to underwrite financially most of the Hungarian construction costs in exchange for future deliveries of electricity to be generated by this project.[33]

Opposition to the project sparked Hungary's first ecological movement. Hungarian ecologists, engineers and economists have long voiced concerns about the project's harmful impacts on the potable water supply, forests, agriculture and settlement systems in both countries, along with the irreparable damage that it would cause to the landscape and natural environment of northwestern Hungary.[34] These various misgivings spawned not only numerous demonstrations throughout the country, but also other forms of protest. These actions delayed the Hungarian start of the project and thus lengthened the proposed timetable.[35] Nevertheless, despite the protests against the project, the Hungarian National Assembly voted in October 1988 to proceed with the project, claiming that abandonment would be too costly. Their decision sparked further massive protests and a 150,000 signature petition campaign calling for a national referendum on the project. A governmental commission of experts formally issued a report on 12 May 1989 recommending that the project be abandoned. The very next day, after a hastily called session of the Hungarian Council of Ministers, the Deputy Foreign Minister announced that work on the 12-year-old project would be suspended. By this time, Czechoslovakia had already completed two-thirds of its share of the project at Gabcikovo.[36]

Hungary's decision to suspend work on its part of the project has been met with rather different reactions in Vienna and Prague. Austria's Foreign Minister, Alois Mock, stated that Hungary's decision would be examined on its economic, technical and ecological merits in a 'spirit of good neighbourliness'. Applauding Hungary's decision, Austrian Greens publicly announced that they would lobby for easy compensation terms for the broken contract. Speaking for the Austrian companies involved in the financing, the Austrian Chamber of Commerce announced that the companies would demand compensation and hold Hungary to the original agreement regarding the delivery of electricity. In order to mitigate the financial losses for the affected Austrian firms, Hungary proposed negotiations to allow them to transfer some of their investments into other joint Austrian-Hungarian projects.[37]

In Prague, however, feelings of 'socialist brotherhood and fraternity' were not very evident. Taking a rigid posture and claiming that the suspension was a politically motivated violation of international law, the Czechoslovakian government demanded that Hungary complete its portion of the project or compensate Czechoslovakia to the tune of US$500 million. To no surprise, several independent Hungarian groups publicly welcomed the suspension.[38] In January 1990, with the political transformations in Eastern Europe well underway, a group of Hungarian environmental protection activists drafted an open letter of appeal to Czech President Vaclav Havel and Czech Foreign Minister Jiri Dienstbier regarding the Gabcikovo-Nagymaros project. Hoping for new discussions between officials of both countries' newly elected governments, they reiterated their environmental concerns about the project and raised the possibility of creating an international nature preserve in the area bordered by the cities of Gyor in Hungary, Bratislava in Czechoslovakia and Vienna in Austria.[39] In mid-May 1990, Hungary was given a list of Austrian demands for damages arising from the halted Nagymaros hydropower plant – totalling 3.8 billion shillings. Signalling a willingness to negotiate a quick resolution of the dispute, the Austrian group indicated a lower figure of 3 billion shillings might be acceptable if agreement could be reached by 30 June 1990. Rejecting the damage claims as too high, Laszlo Udvary, the Hungarian commissioner for the project, went on to state that Hungary wants to cancel its 1988 framework contract with Yugoslavia for the joint construction of four power plants on the Drava River along their common border.[40] At an all-European environmental protection meeting held in Dublin in June 1990, both Prague and Budapest requested that an independent commission of experts from the European Community (EC) be established to study the ecological problems of the entire Danube region.[41] Furthermore, the Czech and Hungarian delegations agreed to settle their disputes surrounding this project based on the findings of this EC commission.[42]

Ironically, while resolution of this volatile dispute has slowly been coming into view, a potentially even more explosive energy-environment conflict may be in the making as a result of the cancellation of these various bilateral international hydroelectric projects. Namely, Electricite de France has offered to build a 900-megaWatt nuclear power plant as part of a US$3 billion investment programme – the Hungarians are seriously examining this offer. The French would provide 70 per cent of the funds, but again, in return for future deliveries of electricity. The planned capacity of this proposed nuclear facility is approximately one-half that of Hungary's only other nuclear plant, which alone satisfies one-third of the Hungarian demand for electricity.[43] The 24-year construction and operational history of this first nuclear power plant, located at Paks, certainly does not augur well for an unruffled future development of a Hungarian nuclear power industry.[44]

Let us leave the discussion about Hungary on a positive note. On 4 June 1990 in Budapest, the senior official of the EC Committee on Environmental Affairs and the Hungarian Minister of the Environment initialled an agreement which will give Hungary 25 million ECU (about US$30 million) by the end of 1990 to carry out 22 energy-saving and environmental projects. This money is to be allocated for such projects as nature conservation (3.2 million ECU), air purification (4.8 million ECU), water purification (3.9 million ECU), waste disposal processing (3.3 million ECU), cleaner power plants and geothermal energy (5.9 million ECU). A further 2 million ECU will be earmarked for the Central and East European Regional Environment Centre, to be established in Hungary at the United States' initiative.[45] This modest sum represents at least a tangible recognition of the international nature and consequences of environmental problems.

The Environmental Deterioration Challenge in Czechoslovakia

Czechoslovakia finds itself in the dubious position of being considered one of the most, if not the most, polluted country in all of Europe – the former East Germany and Poland being the other contenders. The post-war industrialisation plans of the Czechoslovakian regime faithfully followed the Soviet model of emphasising antiquated, inefficient, heavy industries, which required colossal inputs of raw materials and energy and yielded large quantities of polluting residuals – this was much to the detriment of the nation's ecological and economic health. By the mid-1980s, signs began to appear that the Czechoslovakian government was beginning to realise that the economic benefits of additional industrial production were being offset entirely by their associated negative social and environmental costs. In response to this rapidly approaching 'ecological barrier', the Czech and Slovak governments approved, in 1985, a document which for the first time comprehensively dealt with environmental protection. Its measures were incorporated into the eighth (and presumably last) Five Year Plan (1986–1990) and into long-term plans ranging up to the year 2000. This document acknowledged that the intensity of raw material exploitation (per square kilometre) over the past 40 years had grown so fast that it had exceeded the world average by a factor of 15.[46] For example, open cast mining quarries in northern Bohemia as far back as 1980 covered about 11 per cent of the entire area of the region.[47]

Some of the most serious environmental problems and challenges of Czechoslovakia have already been discussed, in particular, the air pollution damages to the forests in the western part of the country, water quality conflicts with Poland and the former East Germany and the protracted Gabcikovo-Nagymaros water project conflict. These three problems, however, barely skim the surface of a very murky situation. Space precludes additional detailed presentation of the ecological disasters in Czechoslovakia, and thus a few examples will have to suffice. For instance, 58 per cent of the forests in the Czech Republic and 35 per cent of those in Slovakia are damaged – severely in much of northern Bohemia. Official data indicate that 70 per cent of the Czechoslovakian rivers are heavily polluted. Approximately one-third of the rivers in the Czech Republic are biologically dead, as are about half of those in Slovakia.[48]

Nonetheless, it would still prove useful for us to review how the Czechoslovakian government of the mid-1980s ranked their environmental deterioration problems in terms of the above cited ecological programme's planned priorities for redressing these problems. Accordingly, they identified the following aspects of the country's environmental deterioration problems for special treatment: (1) air pollution resulting from the burning of high sulphur-lignites; (2) disruption of the water regime due to the declining quality of agricultural land and forestry; (3) water pollution generated from industrial and agricultural activities; (4) reduction of soil fertility; (5) diseased forests and the declining forest soil fertility; (6) loss in the variety of wildlife; and (7) ecological disruption. This ecological programme was to consist of four stages. In the first stage (1986–90), further deterioration was to be stopped, the most seriously impacted areas (northern Bohemia and Prague) were to be improved and more research was to be conducted. In the second stage, the now defunct 1990–1995 plan's target called for at least a 30 per cent reduction in air pollution and environmental improvement throughout the country. Stage three called for a 'substantial' improvement throughout the country. Stage four called for the laudable, but lofty, ill-defined goal of 'restoration of the ecological balance in the nation'.[49]

Obviously, such an ambitious, but felicitous, programme comes with a price tag. While a Czechoslovakian government spokesman confirmed that the government was planning to spend up to Kcs100,000 million by the year 2000, *Hospododarske Noviny* estimated that it would require at least Kcs150,000 million to return the country to the ecological conditions of the early 1970s.[50] Of the government's confirmed sum, approximately Kcs40,000 million would have been required simply to prevent further deterioration, while the remaining Kcs60,000 million would have been required to attain an actual improvement in the environmental quality of the country. At least Kcs17,000 million were allocated for the 1986–1990 Five Year Plan, primarily for air and water purification installations (including some imported technology), assorted equipment for processing and utilising industrial and municipal wastes, and air emissions registration and monitoring equipment.[51] Environmentalists argue that an annual budgetary figure of Kcs9,000 million is necessary, rather than the actual figure of about Kcs3,500 million. Economists counter that such a high expenditure would actually be wasteful, because the overall ecological renewal programme is predicated on two approaches. The first approach involves direct investments and expenditures in pollution prevention and monitoring equipment, while the second approach, or 'intensification process', is indirect. More specifically, the economists foresee a phasing out of dirty and obsolete industries, especially lignite-produced electricity. Implicitly, because much of the larger annual sum that the environmentalists urge spending would be allocated to purification installations at such obsolete facilities, the economists appear correct to consider these expenditures wasteful and counter-productive. On the other hand, except for the profound economic dislocations and plant-closings associated with the conversion to a 'market-based' economy since late 1989, the pace of closing obsolete plants had been going very slowly.[52] To at least partially underwrite this vast financial sum and obtain the requisite pollution abatement technology, Prague looked to the West with some notable success. For example, in June of 1988, Austria's Chancellor Vranitzky stated: '[We] would certainly be willing to consider favorable financing packages. International studies prove that Austria is burdened with the import of pollutants.'[53] Czechoslovakian and (West) German co-operative efforts in the area of environmental protection negotiated prior to the events of 1989 will probably not only survive but expand, especially in the area of nuclear safety.[54]

The role of nuclear power in Czechoslovakia must be mentioned. First, the high levels of air pollution discharged by the lignite-burning conventional power stations in Czechoslovakia help explain why, as previously discussed, Czechoslovakia reacted very negatively to Hungary's decision to suspend construction on its part of the Gabcikovo-Nagymaros hydropower project. Czechoslovakia desperately wants and needs to develop 'clean' alternative sources of much needed electricity. They appear, however, to recognise only two directions – hydro and nuclear power. Despite Chernobyl and nuclear waste and safety problems and issues, the new leaders in Prague seem willing to pursue a nuclear future like their predecessors, except that they are no longer willing to accept the discredited Soviet VVER reactors.[55] Two final items are of interest here. First, on 21 April 1990, Prague announced that Czechoslovakia was suspending construction work on the Gabcikovo project for two years, except for some flood control aspects of the project.[56] And second, in June 1990, the Czechoslovak Institute for Investigation of Public Opinion released the results of an opinion poll on nuclear power, which they had been requested to do by the Ministry of Fuel and Power. According to the poll, 46 per cent of the population supported nuclear power and 41 per cent opposed it.[57] Whether or not this slight plurality will be able to carry the current government into a nuclear future is certainly unclear.

Regardless of the political and economic transformations now underway in Czechoslovakia, the previously discussed plan for ecological renewal elaborated by the discredited regime may not in essence be dead. This is true given the fact that any future Czechoslovakian government will feel itself impelled to attempt to achieve the goals of this ecological restoration programme. This programme may be considered to represent the first candid acceptance of the environmental consequences of 40 years' worth of the now discredited regime's economic policies. In other words, in Czechoslovakia, as elsewhere within the 'former' Soviet bloc of nations, environmental deterioration has played a very central role in challenging the political, economic and even moral legitimacy of Soviet-backed governments. Thus, it is simultaneously understandable and ironic that such novel examples of candour probably rapidly eroded the old regime's legitimacy, rather than strengthened it.

The Environmental Deterioration Challenge in Bulgaria

The 12 most severe Bulgarian pollution problems areas are Sofia, Pernik, Vratsa, Dimitrovgrad, Razlog, Devnya, Pleven, Ruse, Plovdiv, Asenovgrad, Burgas and Kurdzhali. The principal pollutants include dust, nitrogen oxides, lead, ammonia, sulphur dioxide and other hazardous chemicals. The largest polluters are petrochemical factories and coal-burning power plants, especially the Neftochem and Polymet companies in Burgas, and the Pharmacia Company in Stanke Dimitrov.[58] The atmospheres over 19 areas in Bulgaria are admitted to be polluted beyond permissible norms, and 54,000 hectares are to be declared useless from the deposition of heavy metals, such as zinc, lead and cadmium. (This area is presumably in the vicinity of the 'Dimitur Blagoev' Nonferrous Metals Combined Works and the Agria Factory between Plovdiv and Asenovgrad.)[59] Another lead plant at Kurdzhali has also been a major air polluter. Plans have been announced to scale back nonferrous metal production at these plants immediately and to trap gas emissions more efficiently. Furthermore, both are to cease lead production by the end of 1994.[60] Air pollution levels in Dimitrovgrad, reputed to be the fifth worst in Bulgaria, have received considerable recent attention in the National Assembly. In March 1990, the Assembly adopted urgent measures against the three major culprits: the Vulkan Factory, the Dimitrovgrad combined chemical plant and the Maritsa-3 Thermoelectric Power Plant.[61]

Similar to the Green Party in East Germany, the Ecoglasnost group has become a staunch foe of the Bulgarian nuclear power industry. The nuclear plants under construction at Belene and under expansion at Kozloduy have been constant targets.[62] This energy versus environment dilemma is very complex. It is unambiguously and intractably linked with the long-term built-in inertia in Bulgaria's energy balance structure, as depicted in Table 6.4. To crystallise the Bulgarian energy dilemma, one needs merely to be cognizant of the fact that in 1989 the Kozloduy nuclear power plant accounted for 35.6 per cent of the entire electric output of Bulgaria.[63] As of October 1990, the fate of the Belene project was in doubt. News service wire stories from February 1990 indicate that agreement had been reached to halt work on the plant; while later press reports indicate that the agreement is not being observed and that construction work is continuing.[64]

Dozens of media accounts have appeared about the chlorine pollution problems along the Danube, in the cities of Ruse in Bulgaria and Giurgiu in Romania.[65] Here, chronic chlorine leaks and recent earthquake tremors have increased the fears of a large-scale international air-borne chlorine chemical accident at the Giurgiu Chemical Plant, located on the Romanian side of the Danube and across from the Bulgarian city of Ruse. The Bulgarian Ecoglasnost Independent Association has initiated many large-scale protests against the plant and has been able literally to link arms with the Romanian Ecologist Movement Mer over this pollution issue. Bilateral Romanian and Bulgarian efforts to protect the Danube from pollution may well in the long-term be enhanced as a result of the Ruse-Giurgiu pollution controversy.[66]

The Environmental Deterioration Challenge in Romania

Energy related environmental issues have been prominent in Romania.[67] For example, construction of the Cernavoda nuclear power plant has been controversial.[68] Warnings about the depletion of energy resources and their irresponsible exploitation in the past are not uncommon in the Romanian press.[69] More interest in alternative energy sources, such as wind energy, is also finding expression in the newspapers.[70]

Two pollution causes have been highlighted in the media. First, the Romanian press has covered the Giurgiu-Ruse protests and pollution dispute, including the Romanian Ministry of Foreign Affairs' refutation of some Bulgarian charges in this chlorine pollution case.[71] Second, serious carbon sulphide contamination released by the local pulp and paper and artificial fibre combine in the city of Suceava appears to have resulted in at least a temporary closure and thorough inspection of the combine.[72]

As elsewhere in Eastern Europe, the Romanian press contains numerous items discussing the 'Green Movement', the formation of ecological groups and protests, and ecological conferences. Perhaps as a sign of the times, the opening of a private Ecological University in Bucharest was announced on 17 May 1990. From the press conference account, it is difficult to tell just how 'ecological' this stockholder venture is. Supposedly 1200 students can currently be matriculated in its four faculties (medicine, law, engineering and natural sciences).[73] In summary, energy related issues seem to be attracting more Romanian press attention than pollution problems.

The Environmental Deterioration Challenge in Yugoslavia

Finally, a review of recent media reports for Yugoslavia suggests that ethnic and nationality questions receive much more attention than environmental and ecological ones. Nonetheless, the Green Party and Green Movement appear to be active in all republics. At least in some instances, members of the ecology movement may well harbour equally strong ethnic biases, which, unfortunately, would not be very surprising.[74]

Again, similar to other countries of the region, nuclear power seems to be the most controversial ecological issue. The only Yugoslav nuclear plant is located at Krsko, in Slovenia. This 632-kiloWatt plant began producing electricity in the 1970s. Nonetheless, the plant's operation appears to be plagued by minor breakdowns and damage events.[75] In 1984, the Vienna-based International Atomic Energy Agency (IAEA) had made 121 recommendations and proposals for safety work at the plant. In June 1990, IAEA gave the plant a clean bill of health and claimed that demands for closing the plant were unjustified.[76] A mere four days later, however, the new authorities in the Yugoslav Republics of Croatia and Slovenia announced plans to close the Krsko nuclear plant by 1995. The Green Movement in Krsko is reputed to be strong and will not permit the plant to expand its on-site nuclear waste dump, which will reach its full capacity in 1992. Closure would be costly for two reasons. First, there do not appear to be any realistic substitute sources for the plant's annual production of 2.3 billion kWhs. Second, a foreign loan of US$515 million was obtained to construct the plant, and already delayed payments have doubled the amount owed. The current value of the annual electricity production is US$160 million, with an unreported amount available for debt service.[77] While Table 6.4 suggests that Yugoslavia has surplus electricity production, there still are regional energy-environmental problems in the country.

On a positive note, during a speech on 5 June 1990, on the occasion of the World Environment Protection Day, Yugoslav President Borisav Jovic drew special attention to the problems of environmental protection in Yugoslavia and stated that international environmental co-operation with its Danubian and EC neighbours should be a priority.[78] More importantly, this speech foreshadowed a 20 June 1990 announcement by the World Bank – extending a one billion US dollar, 15-year credit (with a possible 5-year grace period and 7.75 per cent interest rate) to Yugoslavia to underwrite half of the costs of a multifaceted environmental protection project. This project's priorities include: (1) protection of the polluted Adriatic Sea and its waterways; (2) securing of the water supplies in the Bosna, Drava, Mura, Sana, Sava and Una rivers in northern and central Yugoslavia; and (3) environmental protection schemes in some industrial zones. Many phases of this project have international environmental significance. Besides the World Bank, Yugoslavia has developed co-operative environmental efforts with (West) Germany, Japan and the United States.[79]

CONCLUSIONS AND UNANSWERED QUESTIONS

From this exploratory overview of the environmental problems and challenges in Eastern Europe, it seems safe to draw some unambiguous conclusions. At the same time, many unanswered questions remain which clearly warrant careful investigation. Unfortunately, these conclusions do not augur well for the environment.

First, the evidence unquestionably points to the unfortunate assessment that environmental deterioration has occurred and is occurring at a truly alarming rate and on a pervasive scale in Eastern Europe. The fundamental reasons for this are both simple and complex. First, the drive for rapid industrial and agricultural development as called for and articulated by the Soviet/Stalinist command-administrative system led to increasing demands for energy production. Second, political decisions to invest disproportionately in dirty 'heavy industries', at the expense of environmentally less damaging light industries and consumer sectors, exacerbated the effects of the ecologically harmful energy extraction and production processes. Third, the natural resource geography of these countries (with somewhat of an exception in the case of Romania) forced them to make excessive use of their collectively most abundant fuels, namely, low-calorie and high-ash lignites and high-sulphur, brown coals, with disastrous air pollution problems as a result. Fourth, the manifold environmental crises in the region are a corollary of the Soviet model's emphasis on the expansion of 'productive' industrial activities, as opposed to investments in 'non-productive' activities. These activities would include the training of ecological and environmental pollution specialists, and research and development in efficient technical pollution prevention and abatement equipment. Fifth, the maxim of 'production maximisation' under the Soviet model of 'extensive' growth policies has clearly demonstrated that it produces deleterious ecological consequences, which are as severe, and in all likelihood, more severe than capitalism's 'profit maximisation' ones. Sixth, the 'former' East-bloc countries' implicit recognition of this situation, and hence, their efforts to respond by transferring to 'intensive growth' and modernisation campaigns, have been doomed from the start by their pricing structures and property rights institutions. In other words, a very important factor in creating the environmental crises in this region has been the irrational administrative price structures in these countries, which have produced both chronically and acutely inefficient resource uses throughout their economies.

Unfortunately, from an environmental perspective, the socio-economic transformations currently underway in Eastern Europe and the Soviet Union pose as many questions as they answer. For example, environmental problems in the West make it pure folly to be sanguine that the various attempts to convert to market-based economies in the region will result in anything more than a palliation of the ecological and resource use problems. The very global nature of environmental problems provides a persuasive argument that similar human economic activities, values and institutional structures and arrangements are the fundamental cause of environmental deterioration. At the same time, the importance of institutions and an educated, ecologically sensitive populous should not be understated. Or, in other words, 20 years of research dealing with environmental problems, mostly those of the Soviet Union, have persuaded this author that clear, firm and guaranteed private property rights are absolutely essential for protecting the environment from degradation. Lamentably, while necessary, rational markets and guaranteed property rights are not by themselves anywhere near sufficient to address and correct the global environmental deterioration problems.

Arguably, the most serious acute and chronic environmental deterioration problems in Eastern Europe are those associated either directly or indirectly with the management of air and water resources. Yet, in fact, it is precisely these two 'fluid resources' which universally take on the difficult management attributes of 'common property resources' and all their associated 'free rider' and 'market failure' problems. At the present, nearly all the countries of Eastern Europe have on paper sound environmental regulations and legislation. Unfortunately, 20 years of the National Environmental Policy Act (NEPA) laws and regulations in the United States have clearly demonstrated that environmental laws and regulations alone simply do not work. The new risk is that as the East European economies become decentralised, their relatively new and weak command regulatory strategies for environmental protection will also be jettisoned without the introduction of market-oriented self-regulatory pollution control structures (for example, effluent charges and taxes). Certainly, but sadly, as these countries look westward for assistance in coping with and solving their manifold environmental deterioration problems, they will find little guidance with regard to efficient market-based, self-regulating pollution control strategies.

Presently, throughout Eastern Europe, two different constellations of reform forces and processes seem to be contending for ascendency to fill the revolutionary vacuum left by the rapid demise of Stalinism. Initially, many of the nascent democratic fronts and groups had strong 'Green' or environmental protection agendas. These 'Green' groups played significant and catalytic roles in bringing down the Stalinist systems. The central espoused goal of these democratic reform movements has been to transform their long subjugated peoples and governments into voting citizens and democratically elected civil societies, respectively. The other vectors of reform, of which Gorbachev represents the prime example, seek a technologic and technocratic revolution which will enable their economies to compete in global markets. Thus far, it seems what modest Western aid has been promised or made available to Eastern Europe has been in the form of financial, technical and managerial assistance, which primarily serve these technologic-technocratic goals. The West seems to be less willing and able to assist in the much more difficult task of developing stable, pluralistic and competitive political institutions and systems. With the reunification of Germany, for example, the numerous neophyte political parties of the former East Germany – including even the 'Green Party' – were quickly and easily overwhelmed by the wealth and organisational power of the established West German parties. While it is true that new market competition will force old inefficient polluting factories in Eastern Europe to close, scarce capital investment funds could also stymie their replacement by non-polluting, new factories. Furthermore, it seems highly likely that environmentalists will be hard pressed to compete with the inevitable fears engendered by rising market prices and the spectre of widespread unemployment. As the Soviet economy continues to unravel, reliable and sufficient oil and natural gas deliveries to Eastern Europe are jeopardised. Thus, the regional energy structure problems of Eastern Europe threaten to beome exacerbated. Concomitantly, as the large state subsidies for utilities to fulfill the 'Communist social contract' of low cost daily living costs for East Europeans disappear, the promises of democracy and a market economy will be seriously tested. Most ominous by far, yet unknown at this writing, is the potential calamity of the recent Gulf War. At the very least, this war threatens aid programmes to Eastern Europe. Finally, inertia in the forms of geography, the human landscapes and infrastructures, and human values, such as nationalism, all constitute formable constraints on what can be done, where, how well and how quickly to improve the environmental conditions of Eastern Europe.

All of these factors lead back to the words written nearly 20 years ago by a fellow geographer, Leslie Dienes, whose personal roots go back to Hungary before 1956. The first part of Dienes' overall assessment of the environmental disruption in Eastern Europe appears to ring as true today as it did in the early 1970s:

> For the near future I am also skeptical concerning the rejection of the ideal of mass consumption and a shift on the people's preference scale toward a simple but ecologically wholesome civilization. The narrow range of consumer goods and sophisticated synthetics in Eastern Europe is viewed not merely as an inconvenience but also as a sign of backwardness, a sore point for these proud nations as infected by the expansionist, acquisitive spirit of Western civilization as the more developed part of the Occidental realm.[80]

Unfortunately, more than 18 years later Dienes' concluding hope 'that even as these nations have telescoped the abuse of environment into a shorter time span, eventually they will resolve the worst aspects of that abuse with less wavering and delay than countries more advanced' remains, not surprisingly, just a hope.[81]

Notes

The author would like to thank Charles Dodd, graduate student, Department of Geography, University of Washington, for his bibliographical assistance and Dr. Morgan Thomas for obtaining graduate student research assistantship support.

1. Alvin Tofler, *Future Shock* (New York: Random House, 1970).
2. See, for example, Marshall I. Goldman, *The Spoils of Progress: Environmental Pollution in the Soviet Union* (Cambridge, MA: MIT Press, 1972); Charles E. Ziegler, *Environmental Policy in the USSR* (Amherst, MA: The University of Massachusetts Press, 1987); Craig ZumBrunnen, 'Soviet Water, Air, and Nature Preservation Problems of the Gorbachev Era and Beyond', in Reiner Weichardt (ed.), *The Soviet Economy: A New Course?: NATO Colloquium* (Brussels: NATO, 1987) pp. 201–28; Craig ZumBrunnen, 'Institutional Reasons for Soviet Water Pollution Problems', *Proceedings of the Association of American Geographers* (Vol. 6, April 1974) pp. 105–8; Craig ZumBrunnen, 'Gorbachev, Economics and the Environment', in *Gorbachev's Economic Plans*, Volume 2 (Washington, DC: US Government Printing Office, 23 November 1987) pp. 397–424; W. A. Douglas Jackson (ed.), *Soviet Resource Management and the Environment* (Columbus, OH: AAASS Press, 1978) pp. 83–104; and Frederick Singleton (ed.), *Environmental Misuse in the Soviet Union* (New York: Praeger Publishers, 1976).
3. Ivan Volgyes (ed.), *Environmental Deterioration in the Soviet Union and Eastern Europe* (New York: Praeger Publishers, 1974).
4. Leslie Dienes, 'Environmental Disruption in Eastern Europe', in Volgyes, *op. cit.*, in note 3, pp. 141–58.
5. Joan DeBardeleben, *The Environment and Marxism-Leninism: The Soviet and East German Experience* (Boulder, CO: Westview Press, 1985).
6. Barbara Jancar, *Environmental Management in the Soviet Union and Yugoslavia: Structure and Regulation in Federal Communist States* (Durham, NC: Duke University Press, 1987).
7. George W. Hoffman, 'Eastern Europe', in George W. Hoffman (ed.), *Europe in the 1990s: A Geographic Analysis* (New York: John Wiley & Sons, 1989) pp. 602–8.
8. *Ibid.*, pp. 608–29.
9. IIASA staff, 'The Price of Pollution', *Options* (September 1990) p. 408.

10. G. H. M. Krause, *et al.*, *Forest Decline in Europe: Possible Causes and Etiology* (Toronto, Canada: paper presented at the International Symposium on Acid Precipitation, September 1985); and figures from *Allgemeine Forst Zeltschrift* (No. 41, 1986).
11. Nico van Breeman, 'Acidification and Decline of Central European Forests', *Nature* (2 May 1985); and Tomas Paces, 'Sources of Acidification in Central Europe Estimated from Elemental Budgets in Small Basins', *Nature* (2 May 1985).
12. IIASA staff, *op. cit.*, in note 9, p. 6.
13. Paces, *op. cit.*, in note 11.
14. Andrew Csepel, 'Czechs and the Ecological Balance', *New Scientist* (27 September 1984).
15. '"Environmental" Protection Accord with Finland', *Foreign Broadcast Information Service Daily Reports: Eastern Europe 90–090 (FBIS-EEU-90–090)* (5 May 1990) p. 37.
16. *Ibid.*, p. 37.
17. 'Environment Protection Agreement with Sweden', *FBIS-EEU-90–101* (24 May 1990) pp. 65–6.
18. Claus-Juergen Goepfert, 'The Hidden Camera of GDR Ecologist', *Frankfurter Rundschau* (7 November 1988) p. 5.
19. DeBardeleben, *op. cit.*, in note 5, p. 36.
20. *Ibid.*, p. 37.
21. See, for example, 'Government Discusses 1975 Nuclear Accident', *FBIS-EEU-90–015* (23 January 1990) pp. 34–5; 'Alles Starben', *Der Spiegel*, (22 January 1990) pp. 85–7; 'Nuclear Power Plant Expansion Problems Noted', *FBIS-EEU-90–023*, (2 February 1990) pp. 46–7; and 'Accusations Against Nuclear Plant "Untenable"', *FBIS-EEU-90–042* (2 March 1990) p. 41.
22. See, for example, 'Greens, Democracy Now Nominate Candidates', *FBIS-EEU-90–043* (5 March 1990) p. 33; 'Election Programs of IWU, Green Party', *FBIS-EEU-90–044* (6 March 1990) p. 28; 'UFV in Protest Ends Alliance with Green Party', *FBIS-EEU-90–058* (26 March 1990) p. 25; and 'Women's Union Breaks Coalition with Greens', *FBIS-EEU-90–059* (27 March 1990) p. 25.
23. 'Environmental Official on Energy Sector Reshaping', *FBIS-EEU-90–117* (18 June 1990) p. 41.
24. 'Nuclear Power Plant Construction Debated', *FBIS-EEU-88–194* (6 October 1988) p. 36.
25. 'Demonstration Against Nuclear Power Plant', *FBIS-EEU-90–080* (25 April 1990) p. 56.
26. 'Demonstrators Protest Nuclear Plant', *FBIS-EEU-90–116* (15 June 1990) p. 42.
27. 'Sejm on CSSR Dispute, Nuclear Power', *FBIS-EEU-88–222* (17 November 1988) p. 47.
28. 'Czech Premier Petr Pithart Winds up Visit Signs Environmental Agreement', *FBIS-EEU-90–083* (30 April 1990) p. 41.

29. 'Talks with GDR on Border Waters Protection', *FBIS-EEU-909–020* (30 January 1990) p. 64.
30. Roman Stefanowski, 'Poland Not To Be a Dumping Ground for Ecological Waste?', *Radio Free Europe Research: Polish Situation Report/3* (February 1989) pp. 17–18.
31. For information about nuclear power issues in Hungary, see 'Stricter Nuclear Regulations', *FBIS-EEU-88–187* (27 September 1988) p. 25; and Steven Koppany, 'Hungary's First Nuclear Power Plant A Monument to Inefficiency', *Radio Free Europe Research: Hungarian Situation Report/5* (25 March 1986) pp. 19–23.
32. Hoffman, *op. cit.*, in note 7, p. 607.
33. Edith Markos-Oltay, 'Government Suspends Work on Nagymaros Hydroelectric Project', *Hungarian SR/8* (30 May 1989) pp. 21–4.
34. *Ibid.*, pp. 21–2.
35. See, for example, 'Closing Document on Danube Barrages Prepared', *FBIS-EEU-88–173* (7 September 1988) pp. 28–9.
36. Markos-Oltay, *op. cit.*, in note 33, pp. 21–2.
37. *Ibid.*, p. 22.
38. *Ibid.*, p. 23; and J. Tagliabue, 'Hungary Quits Danube Project, Angering Prague', *The New York Times*, 19 May 1989.
39. 'Environmentalists Appeal to Havel, Dienstbier', *FBIS-EEU-90–010* (16 January 1990) p. 69.
40. 'Austrian Demands for Nagymaros Damages Rejected', *FBIS-EEU-90–112* (11 June 1990) p. 58.
41. 'EC Asked to Mediate in Power Plant Dispute', *FBIS-EEU-90–118* (19 June 1990) p. 38.
42. 'Environmental Protection Body Views Danube Project', *FBIS-EEU-90–120* (21 June 1990) p. 37.
43. *Op. cit.*, in note 40, p. 58.
44. Koppany, *op. cit.*, in note 31, pp. 19–23.
45. 'Agreement with EC Signed on Environmental Aid', *FBIS-EEU-90–109* (6 June 1990) p. 40.
46. 'Approaching the "Ecological Barrier"', *Radio Free Europe Research: Czechoslovak Situation Report/9* (18 June 1986) pp. 23–4.
47. 'How the Party (Mis)manages Northern Bohemia', *Czechoslovak SR/10* (11 July 1986) p. 28.
48. 'The Destruction of the Environment Continues', *Czechoslovak SR/6* (30 March 1989) pp. 13–14.
49. The final version of this ecological programme was published in *Hospodarske Noviny* (No. 20, May 1986, supplement).
50. *Hospodarske Noviny* (No. 42, 25 October 1985) p. 3.
51. *Op. cit.*, in note 46, p. 24.
52. 'A Difficult Ecological Recovery', *Czechoslovak SR/14* (22 October 1986) pp. 24–5.
53. Kurt Seinitz, 'Austria Willing to Prefinance CSSR Environmental Protection', *FBIS-EEU-88–126* (30 June 1988) p. 13.

54. 'Ecological Cooperation with the FRG', *Czechoslovak SR/14* (25 September 1987) pp. 21–4; and 'Nuclear Safety Agreement Signed with FRG', *FBIS-EEU-90–105* (31 May 1990) p. 11.
55. 'USSR No Longer Responsible for Nuclear Safety', *FBIS-EEU-90–092* (11 May 1990) p. 11; 'Changes in Nuclear Power Concept Planned', *FBIS-EEU-90–096* (17 May 1990) pp. 17–18; and 'The Energy Outlook to the 21st Century', *Czechoslovak SR/13* (11 September 1987) pp. 19–24.
56. 'Gabcikovo Suspended for Two Years', *FBIS-EEU-90–081* (26 April 1990) p. 18.
57. 'Public Opinion Survey on Nuclear Plants', *FBIS-EEU-90–114* (13 June 1990) pp. 28–9.
58. '"Urgent Measures" Adopted', *FBIS-EEU-90–117* (18 June 1990) p. 11.
59. 'Environmental Ministry's Activities Criticized', *FBIS-EEU-90–083* (30 April 1990) p. 11; and 'Lukanov Meets Industrial Polluters', *FBIS-EEU-90–091* (10 May 1990) p. 8.
60. 'Ecology, Health Measures for Lead Plant', *FBIS-EEU-90–056* (22 March 1990) p. 12.
61. 'Debates Ecological Problems', *FBIS-EEU-90–063* (2 April 1990) p. 1.
62. See, for example, 'Ecoglasnost Opposes Nuclear Plant Construction', *FBIS-EEU-90–014* (22 January 1990) pp. 15–16; 'Work Begins at Kozloduy Plant', *FBIS-EEU-90–014* (22 January 1990) p. 16; Krasimir Tsigularov, 'The Energy Industry: Time For Doubts', *FBIS-EEU-90–017* (25 January 1990) pp. 15–16, translated from *Rabotnichesko Delo* (23 January 1990) pp. 1 and 4; 'Ecoglasnost Protests Belene Nuclear Plant', *FBIS-EEU-90–017* (25 January 1990) p. 16; and 'Svishtov Strike Against Belene Nuclear Plant', *FBIS-EEU-90–018* (26 January 1990) p. 11.
63. 'Ecoglasnost Opposes Nuclear Plant Construction', *op. cit.*, in note 62, p. 16.
64. 'Protocol Signed on Nuclear Power Plant', *FBIS-EEU-90–055* (21 March 1990) p. 8; and 'Ecological Protests Belene Nuclear Plant', *FBIS-EEU-90–110* (7 June 1990) p. 10.
65. See, for example, 'Ecoglasnost Threatens Protest over Romanian Plant', *FBIS-EEU-90–106* (1 June 1990) p. 5; 'Ecoglasnost Delegation Holds Talks in Giurgiu', *FBIS-EEU-90–108* (5 June 1990) pp. 10–11; 'Government "Urgent Measures" on Ruse Pollution', *FBIS-EEU-90–110* (7 June 1990) p. 9; 'More Chlorine Emissions Reported from Ruse', *FBIS-EEU-90–111* (8 June 1990) p. 15; and 'Ruse-Giurgiu Pollution Also Discussed', *FBIS-EEU-90–117* (18 June 1990) p. 11.
66. 'Danube Pollution Discussed in Bucharest', *FBIS-EEU-90–117* (18 June 1990) p. 10.
67. See, for example, 'Nuclear Hydroelectric Power Potential Assessed', *FBIS-EEU-90–045* (7 March 1990) pp. 71–2; 'Romania Discusses

Energy Matters with Experts, *FBIS-EEU-90–045* (7 March 1990) p. 72; and 'Electric Power Minister Speaks on Energy Needs', *FBIS-EEU-90–053* (19 March 1990) p. 77.

68. 'CAE Cernavoda – A Fundamental Overhaul', *FBIS-EEU-90–016* (24 January 1990) p. 67; and 'Stage of Construction Works on the Nuclear Power Plant at Cernavoda', *FBIS-EEU-90–014* (22 January 1990) p. 89.
69. See, for example, 'Warning Against Depleting Energy Resources', *FBIS-EEU-90–060* (28 March 1990) p. 64.
70. '"Wind Energy Association" Forms in Bucharest', *FBIS-EEU-90–064* (3 April 1990) p. 62.
71. See, for example, 'Bulgarian Protests Against Pollution Noted', *FBIS-EEU-90–106* (1 June 1990) p. 55; 'Joint Bulgarian Air Pollution Meeting in Giurgiu', *FBIS-EEU-90–066* (5 April 1990) p. 58; and 'Bulgarian Changes on Giurgiu Pollution Refuted', *FBIS-EEU-90–111* (8 June 1990) p. 49.
72. 'Suceava Pollution Cause for Combine to Close', *FBIS-EEU-89–022* (1 February 1989) p. 65.
73. 'Private Ecological University Opens in Bucharest', *FBIS-EEU-90–099* (22 May 1990) p. 52.
74. 'Ecology Movement Founded in Montenegro', *FBIS-EEU-90–098* (21 May 1990) p. 77.
75. 'Krsko Nuclear Power Plant Shut Down Due to Damage', *FBIS-EEU-88–233* (5 December 1988) p. 60.
76. 'Krsko Nuclear Plant Deemed "Safe" by IAEA', *FBIS-EEU-90–109* (6 June 1990) p. 56.
77. 'To Close by 1995', *FBIS-EEU-90–109* (6 June 1990) p. 56.
78. 'Jovic Speech Marks World Environment Day', *FBIS-EEU-90–108* (5 June 1990) p. 81.
79. 'World Bank Grants US$1 Billion Ecological Credit', *FBIS-EEU-90–120* (21 June 1990) p. 74.
80. Dienes, *op. cit.*, in note 4, p. 154.
81. *Ibid.*, p. 154.

7 Institutional and Legal Responses to Global Environmental Change

Glen Plant

INTRODUCTION

The international community is encountering significant challenges as a consequence of global environmental change, which is apparently resulting from modern human activities, in particular the industrial production of various gases. The fundamental question facing the international community is whether the global community should seek to administer and co-ordinate its responses through existing structures, or establish one or a number of new international institutions.

The political importance of the negotiations[1] is such that they are likely to lead to some degree of institutional reform in relation to the treatment of environmental issues in general. The international agenda on issues of global environmental change is dominated by concerns over both the potential depletion of the earth's protective ozone layer and the possible worldwide climatic warming, which may result from increased concentration of 'greenhouse' gases in the atmosphere. Though each potential development is of vital concern for the global community, the institutional and legal challenges of the latter are much greater than the former.[2] Thus, this chapter will focus upon institutional and legal responses to global warming.

It is my purpose in this chapter to examine the role that existing international organisations have played both in the current movement towards a global climate change convention and in the regional and national initiatives that have accompanied this movement, and which may ultimately have to serve as substitutes for such a convention. In the light of this examination, I will consider not only a number of possible reforms to the existing institutions, but also the feasibility of creating new ones. I will, moreover, attempt briefly to evaluate the impact that the most probable reforms will be likely to have upon the future formation of international law, particularly in the environmental field.

The Report of the World Commission on Environment and Development

(WCED), finalised in 1987, suggested that the challenges of global warming

> are both interdependent and integrated, requiring comprehensive approaches and popular participation . . . Yet most of the institutions facing those challenges tend to be independent, fragmented, working to relatively narrow mandates with closed decision processes.[3]

There is a fear that an international institutional 'organisational crisis' will develop in tandem with the aggravation of environmental threats. This could result in a widening gap between international environmental objectives and the capacity of organisations to deal with those objectives.[4]

A number of states have, therefore, chosen to propose the establishment of new organisations. At The Hague Ministerial Conference, which was attended by 24 selected states in March 1989, France, Norway and the Netherlands proposed the creation of a new international organisation with substantial powers.[5] Although the idea was considered by the participating states, it was adopted in the Final Declaration of The Hague in a less extreme form.[6]

Other states, including the United Kingdom, might well take the view that eventually an international mechanism to enforce international environmental obligations, such as that proposed in the draft Hague Declaration, might prove necessary; at present, however, they regard these proposals as premature at best. This view is perhaps best expressed by Sir Crispin Tickell – the then British Permanent Representative to the UN – in his statement before the UN Economic and Social Council (ECOSOC) on 9 May 1989:

> Time does not allow us the luxury of trying to create something new . . . We have institutions enough already. The challenge we face is how to make the best use of them, and if necessary adapt them to changing circumstances.[7]

PARTICIPATION OF EXISTING INSTITUTIONS

UN Programmes and Specialised Agencies

Since the Stockholm Conference on the Human Environment in 1972, the primary role of safeguarding the environment as a whole has fallen upon the United Nations and, in particular, the United Nations Environment Programme (UNEP). It has been suggested that the principal accomplishment of that conference was the 'official recognition of the environment as a subject of general international concern and the institutionalization of that concept in UNEP'.[8] Although other international and specialised agencies – both UN and non-UN – play a role in the management of the environment, it is UNEP which is specifically assigned 'a catalytic and coordinating role' within the UN system.[9]

It is unnecessary for the purposes of this chapter to examine in detail the full scope of UNEP's activities and its committee structure. It is sufficient to note two matters concerning its relationship with other intergovernmental bodies. The first is the vertical chain of accountability. UNEP is not a specialised agency, but merely a programme of the United Nations General Assembly (UNGA) with a Governing Council elected by and responsible to UNGA. The reports of UNEP's Governing Council are submitted first to ECOSOC and then to UNGA, where they are examined in the Second Committee before being approved in the plenary sessions of UNGA.[10] Thus its reports are examined in three stages.

The second is the horizontal arrangement of formal and informal co-ordinating mechanisms with UN and specialised agencies. As far as formal relations are concerned, UNEP administers, through its Administrative Committee on Co-ordination (ACC), the Secretariats of the two principal co-ordinating committees: the Designated Officials for Environmental Matters of the United Nations (DOEM) and the Committee of International Development Institutions on the Environment (CIDIE).[11] UNEP, the World Bank (IBRD), the UN Development Programme (UNDP) and 11 other intergovernmental financial institutions are represented on CIDIE. It is noteworthy that the World Bank has declined to participate in DOEM meetings, although it is somewhat reassuring that it has recently announced a decision to set up, in co-operation with UNEP and UNDP, a Global Environmental Facility. From this fund, scheduled to commence in July 1991 for an initial period of three years, developing states will be able to draw financial aid to offset some of the negative environmental impacts of developmental projects.[12] The World Bank has also begun to consider the environmental impact of proposed projects for which it provides loans from its existing facilities.

UNEP also co-operates informally with a large number of UN specialised agencies and other institutions whose work has environmental aspects. In the context of global climate change, the most important have been the World Meteorological Organization (WMO), the Food and Agriculture Organization (FAO), the United Nations Educational,

Scientific and Cultural Organization (UNESCO) and the United Nations Development Programme (UNDP).[13] There is, of course, little scope for formal reform in relation to these informal links, but an enhanced importance or status for UNEP is likely to reduce the residual resistance of such organisations to the growing incursion of a relatively small and new organisation – UNEP – into matters formerly and exclusively within their own bailiwicks: this may result in more fruitful informal co-operation in future.[14]

These UN institutions have established a large number of joint programmes concerned with data collection and research into global climate change and the formulation of policy responses. Two of the most important are worth noting. The first is the World Climate Programme (WCP), a scientific programme with several sub-components,[15] in which the non-governmental scientific body, the International Council of Scientific Unions (ICSU), also participates. This compliments the work of the International Geosphere-Biosphere Programme (IGBP), itself initiated by ICSU.[16]

The second is the Intergovernmental Panel on Climate Change (IPCC). Created by UNEP and WMO, it first met in November 1988 in Geneva. The IPCC was subsequently charged by UNGA with the task of preparing a comprehensive review of options and recommendations in response to global warming.[17] The IPCC's three working groups (WG I, WG II and WG III) completed this task, as scheduled, in advance of both the Second World Climate Conference in Geneva in November 1990 and the 45th Session of UNGA.[18] Their work and the results of that conference are to be taken into account in the negotiation of a global framework climate change convention, which, it is hoped, will be ready to be opened for signature at the World Conference on Environment and Development (the 1992 Conference), to be held in Rio de Janeiro in June 1992.[19]

Finally, mention should be made of recent changes in some of the intergovernmental financial institutions that have occurred since account of their role was given by WCED in 1987.[20] (This account was rather negative despite their earlier participation in the 1980 Declaration on Environmental Policies and Procedures Relating to Economic Development.)[21] In the late 1980s, largely as a result of increased public awareness and outcry at the negative environmental impacts of development loans made by those institutions (in particular, the World Bank loans to Brazil), they began to increase their staff and create new environmental departments within their secretariats and also to reflect environmental safeguards in their loan agreements.[22]

Intergovernmental Institutions of the Developed World

(a) The Organization for Economic Cooperation and Development (OECD)
Although the OECD's primary concern is with issues of economic development and international trade, it established an Environment Committee in 1970 to provide a forum for member countries to discuss environmental problems of common concern and to propose recommendations for national policies.[23] This committee co-operates with other international bodies, such as the Commission of the European Communities, which attends its meetings as an observer.

(b) The UN Economic Commission for Europe (UNECE)
The UNECE has been involved in environmental issues since at least 1971, when it organised a symposium on Problems Relating to the Environment in Prague, Czechoslovakia.[24] In May 1990, it convened a meeting of 32 European and North American states at Ministerial level in Bergen, Norway to consider further the issues raised by the WCED and to establish 'national strategies and/or targets and schedules following the report of the IPCC and no later than the start of negotiations of a framework Convention'.[25] These countries agreed to commit themselves to the adoption of national policies on response to climate change by the end of 1990, that is well in advance of the 1992 Conference. Many have done so, and some, like Denmark and Germany, have adopted particularly stringent targets.[26] Unfortunately, concealed in the apparent success of this approach is a danger that even the industrialised nations might be unable to agree upon meaningful common standards, whatever form these ultimately take.[27] The more industrialised states might find themselves with more to surrender than less industrialised nations, such as the Scandinavian states.

(c) The Group of Seven (G7)
The communiques of the Group of Seven (the seven richest industrialised nations which hold annual summit meetings) in Paris in July 1989 and in Houston, Texas in June 1990 made reference to the need for a global framework convention on climate change.

(d) The European Communities (EC)
The EC has for some years been concerned with the question of emission standards in relation to nitrogen and sulphur oxides, largely because of particular regional concerns about the acid precipitation effects of these gases.[28] Nevertheless, more recently, for instance at the Rhodes European

Council of December 1988, the EC has become more concerned with emissions of the other important greenhouse gases, especially carbon dioxide. In May 1989, for example, the Environment Commissioner first mooted proposals for a carbon tax, which are presently under active consideration.[29] In June 1989, the Council of Ministers passed a resolution that underlined 'the global dimensions of the greenhouse effect and the need for the Community and the Member States to play their full part in the definition and implementation of a global response to the problem'.[30] Nevertheless, it also welcomed the EC Commission's initiative in launching a substantial policy-options study programme in parallel with that of the IPCC. It seems imperative that the EC will need to develop a common policy on the issue. At present, efforts appear to be directed towards the idea of harmonised, community-wide environmental taxation, a concept to which the United Kingdom Government, at least, is opposed.

The most significant contributions of these organisations to the climate change debate are perhaps the provision of initiatives and the setting of standards on a regional or national basis in advance of the finalisation of negotiations on a global climate change convention. These contributions are likely both to advance thinking on the possible content of that convention and to provide an alternative 'mini regime' system based on participation of the present key greenhouse gas-producing states in the event that the hard policy consequences tied up with a global package cannot be agreed at the global level.[31]

Intergovernmental Organisations of Developing Countries and Other Regions

Few other intergovernmental organisations appear to have taken any significant steps in the movement towards a global regime to combat climate change. These steps have essentially been limited to declarations of concern, such as the Caracas Declaration of the Special Ministerial Meeting of the Group of 77 in June 1989,[32] the Declaration of the Ninth Non-Aligned Summit in Belgrade in September 1989[33] and the Langkawi Declaration of the Commonwealth Heads of Government in October 1989.[34] The Commonwealth has at least established a study group on the subject.

Nongovernmental Organisations

A large number of nongovernmental or scientific organisations and pressure groups, both international and national, have played roles in the development of responses to global warming to date.

(a) International Council of Scientific Unions (ICSU)

The ICSU has been the most significant nongovernmental organisation in the climate change issue. Besides its participation in the WCP and IGBP programmes,[35] ICSU also organised, together with WMO and UNEP, the 1985 Villach scientific conference on greenhouse gases,[36] which called for policy responses to climate change from governments and other scientific organisations in collaboration with them and which set up an Advisory Group on Greenhouse Gases.[37] This conference led to a two-stage workshop in 1987 in Villach, Austria and Bellagio, Italy.[38]

(b) World Resources Institute

The World Resources Institute has participated in at least two programmes in conjunction with intergovernmental organisations (*infra*).

(c) International Union for Conservation of Nature and Natural Resources (IUCN)

Although it is a body primarily concerned with conservation issues, it seems possible that the IUCN will increase its involvement in the climate change issue. In 1988, its General Assembly requested its Director-General, as part of the 1988–90 programme, both to 'cooperate with UNEP, WMO and other organizations undertaking basic analysis and modelling of the "greenhouse effect"' and to 'take steps, in cooperation with IUCN members and others, to promote the actions identified as necessary to minimize global warming of the atmosphere and its harmful effects on the environment'.[39]

Besides these three organisations, a number of new organisations, intended to link intergovernmental and private expertise, have been established. In 1984, the World Industrial Conference on Environmental Management, organised by UNEP and the International Chamber of Commerce, brought together leaders of industry, governments and NGOs and concluded 15 major recommendations. The Aspen Institute, established in 1990 by a Michigan corporation, UNEP and NASA, aims to bring together scientists to exchange ideas on the environment.[40]

CURRENT DEVELOPMENTS: THE CONTRIBUTION OF EXISTING INSTITUTIONS

The responses of existing institutions to climate change have gone through several stages. The first was scientific. Although it was postulated as early as the last century that changes in the concentration of certain gases in the atmosphere would lead to global warming,[41] it was not until the 1970s that scientific research began to indicate not only the depletion of the ozone layer, but also the potential occurrence of enhanced global warming. The earliest international response was thus in the form of improving scientific programmes, for example, through the establishment of the WCP in 1979 and the Villach-Bellagio conferences already mentioned.[42] It was largely as a result of these efforts that Working Group I of the IPCC was able to conclude that the balance of probabilities was in favour of the existence of enhanced global warming,[43] and that it became evident that a global approach to the problem was desirable.

The second stage was the movement towards confirming the scientific evidence and potential impact of the effect and towards formulating various policy options. Thus, following the publication of the WCED report, a WMO/UNEP-sponsored conference took place in Toronto, Canada on 'The Changing Atmosphere: Implications for Global Security'.[44] In the keynote address, Canadian Prime Minister Brian Mulroney called for a concerted international effort to develop by 1992 an international accord or elements thereof for the protection of the atmosphere.

The conference led to a Meeting of Legal and Policy Experts in Ottawa, Canada in February 1989, which was attended by 80 experts from governments, international organisations (including UNEP and WMO) and universities. Its terms of reference included the task of developing

> the legal and institutional framework for dealing with existing and emerging atmospheric problems and, where possible, to agree on principles that might form the basis of an umbrella framework convention for the protection of the atmosphere and for a convention on climate change.[45]

The meeting suggested that progress be made on a general atmospheric convention as well as a climate change convention, despite the advice of Mostafa Tolba, UNEP's Executive-Director, that only the latter should be pursued at present, especially in view of the present overcrowded international agenda on environmental issues.[46] Nevertheless, the Statement of the meeting appears to have placed greater emphasis on climate change. In addition to proposing such innovations as the establishment of an Experts' Panel on global warming issues and a World Climate Fund, the statement suggested that the 'framework' approach of the Vienna Convention on the Protection of the Ozone Layer (the Vienna Convention)[47] be used as a model for a future climate change treaty.

A choice had to be made in effect between two different forms of agreement: a comprehensive 'umbrella' convention, which deals at once with the broadest possible range of matters and essentially leaves only technical details to be supplied later, rather along the lines of the Law of the Sea Convention; or a 'framework convention', which, like the Vienna Convention, sets out the measures necessary in principle and provides a general framework for future protocols on specific greenhouse gases. Given the present uncertain state of scientific knowledge of greenhouse gases and their interaction, the latter quickly became the preferred alternative.

The Vienna Convention is a useful model, not only because it and its Protocol, as amended,[48] deal with chlorofluorocarbons (CFCs), which are major contributors to the greenhouse effect, but also because this is the first international treaty on the environment to include financial aid and procedures for transfer of technology to less developed countries.[49] Both environmentalists and foreign policy experts consider appropriate transfer mechanisms to be similarly a critical requirement for the success of any international agreement on global warming. It also demonstrates the flexibility of the framework approach. The terms of the Protocol, originally opened for signature in 1987, have been tightened as scientific knowledge has progressed; it is to be reviewed again in 1992. On the other hand, the ozone model is in other respects a discouraging one, because it has set relatively low standards. Of course, ozone layer protection is a much less complex issue than global warming. Thus, the value of the model and the chances of success in relation to the latter might prove to be even more limited than they are in relation to the former.

Besides the establishment of the IPCC in November 1988, the importance of climate change on UNEP's agenda was increasing at this time.[50] From the viewpoint of institutional change, Working Group III on Response Strategies, chaired by a US expert, is the most significant IPCC working group. It established a steering committee, with four sub-committees, to discuss different sectoral problems and strategies to limit or adapt to climate change. It then divided its work into two 'Tasks': Task A, 'Emission Scenarios' and Task B, 'Implementation Mechanisms'.[51] Task B was divided into five topic groups, of which one concerned legal and institutional measures; responsibility for this topic was assigned to the UK, Canada and Malta. In addition to promoting relevant resolutions in UNGA,[52] these governments have drafted possible framework conventions for discussion.

Working Group III concluded that, given the emergence of a consensus at the 44th Session of UNGA in favour of a global framework convention:

> the measures [it recommends] require a high degree of international cooperation with due respect for national sovereignty of states. The

> international negotiation on a framework convention should start as quickly as possible after the completion of the IPCC First Assessment Report.[53]

In August 1990, representatives from 75 nations met in Sundsvall, Sweden at the fourth meeting of the IPCC and approved its first assessment report. Unfortunately, it is possible to have some reservations about the strength of commitment of certain industrialised countries to the negotiation of effective standards. At the Noordwijk Ministerial Conference on Atmosphere, Pollution and Climate Change, which was held in November 1989 and attended by 70 states, a number of countries (including the US, UK, Japan and, surprisingly, West Germany) lined up behind proposals for disappointingly lenient standards.[54]

Of the three IPCC working groups, moreover, Working Group III has been criticised for being the most conservative. The only protocol to the proposed framework convention that it suggested investigating – in terms of feasibility in the short term – was a protocol on deforestation. This might well have been influenced by the pre-existence of proposals for an initiative on deforestation. Such a protocol is to be regarded with some scepticism for fear that it might be used as an excuse to divert attention away from more serious factors contributing to global warming.

It is possible, therefore, that the third stage, the commencement of negotiation of a global climate change convention, will not be accompanied, at least not in the short term, by any truly effective protocols on individual greenhouse gases. Nevertheless, the third stage is well under way. The first meeting of the preparatory commission for the 1992 Conference was held in March 1990.[55] Its first substantive session took place in August 1990, with UNEP playing a co-ordinating role. A further meeting was planned for March 1991, and the suggestion has been made that further preparatory meetings be held at three-monthly intervals of two weeks' duration, right up to the 1992 Conference.

POSSIBLE REFORMS TO EXISTING INSTITUTIONS AND ESTABLISHMENT OF NEW INSTITUTIONS

The analysis of possible reforms must, therefore, proceed in the first instance upon the assumption that a global framework convention will be successfully negotiated by 1992, together with perhaps only a resolution of the Conference in favour of the rapid conclusion of necessary protocols, or at most, a single protocol on deforestation. The alternative scenario must be the failure to achieve global agreement and the establishment of a mini treaty regime by the OECD or another similar body.

New Institutions

(a) Convention Institutions

The most obvious result of this negotiation is likely to be the establishment of a Secretariat to assist the negotiation. This, it appears, will be supplied by the existing UNEP and WMO Secretariats. It may well be that once the convention is in place, a small Secretariat to oversee its implementation will need to be created. It is likely that rather than the IPCC being transferred into a Secretariat, the new Secretariat will be supplied initially by the UNEP Secretariat, which has experience of establishing convention-specific secretariat groups or implants. Alternatively, it may be supplied by WMO or even established *de novo*. The UK has suggested transforming the IPCC, whose task is not expected to be completed by 1992, into an independent Intergovernmental Commission with a supporting role for the Convention rather than any central role.[56] The question would then arise whether this body should report to UNEP and WMO or to a more central UN organ instead. The choice will, to some extent, be affected by changes in the status of UNEP presently being mooted (*infra*).

The question of funding would have to be addressed. Judging from past experience – for example, the 1973 Convention on International Trade in Endangered Species and the 1978 Convention on Conservation of Migratory Species – the most likely course to be taken would be the provision of finance from UNEP's Environmental Fund, superseded over a period of time by a Trust Fund financed by the States Party. In this case, however, finance might come immediately from any World Climate Fund which is set up.

The convention is likely to involve an unprecedented degree of international co-operation concerning the control of multi-media activities; it is also likely to result in relatively high degrees of friction over its proper implementation. Thus, it might well be in relation to this very convention that the international community will first come to accept the need for an institution with real teeth, that is, an institution with the ability to impose onerous standards and to enforce them effectively. It is difficult to predict how such an institution would evolve, and in any event, because several protocols would first need to be in operation, its very existence is probably only a long-term possibility. It seems unlikely that states would accept compulsory and binding dispute-resolution procedures, such as those found in the Law of the Sea Convention, because of the exceptional economic importance of the issues involved. They might, however, be prepared to envisage a less binding form of enforcement or dispute-settlement machinery, such as something along the lines of the GATT procedures.

Parallels are being increasingly drawn between environmental and human rights issues. Given that there are strong arguments in favour of an existing human right to a decent environment, such new institutions might well follow similar patterns to existing human rights institutions. These would be more likely to follow the UN system of Commission and Committee to oversee an international instrument, rather than the more centralised and binding procedures of the European Convention on Human Rights. In this connection, of course, one might refer to concepts of intergenerational equity that are increasingly being expressed.[57] The suggestion has been made that an international environmental ombudsman might be created to represent the interests of future generations, presumably in relation to administrative and judicial decisions likely to have serious environmental costs for future generations.[58] There appears to be no good reason why these procedures should not work, given the requisite political will, but they are likely to be matters for discussion in the longer term after a convention and network of protocols imposing specific standards are in place. Clearly, any such new institutions would need to be placed under the supervision of an existing UN organ, preferably UNGA.

What remains to be seen is how specific controls on greenhouse gases are to be embodied. For example, in relation to carbon dioxide – objectively the primary candidate for the first protocol on a specific greenhouse gas – it is unclear at present whether controls on its overproduction could be best effected by emissions controls, a carbon tax, a purchased quota system, an 'insurance' fund, development aid, a trust fund or a combination of these and perhaps other methods. If a fund approach were preferred, a new financial body might be necessary to administer it, assuming that there might well be objections to the traditional international financial institutions' playing a role; and, if a tax were preferred, a system of collection and inspection, presumably to be conducted by a new or expanded institution, would be required. It is perhaps significant in this respect that the Montreal Protocol fund has been established by the Protocol's Executive Committee independently of the World Bank's new Global Environmental Facility (*supra*). This is presumably to avoid elements of 'conditionality', as opposed to 'additionality', creeping into the administration of such a fund.

(b) Non-Convention-Related Institutions

It is difficult to predict what entirely new institutions might be created beyond those immediately dependent upon the terms of the new global convention. What can be suggested is that some new organisations might arise in relation to matters that, like ozone layer protection, span environmental concerns going beyond global warming *simpliciter*. Thus, for example, a new world forestry body, the 'Global Forestry Organisation', has been suggested by a group of forestry experts critical of many aspects of the existing Tropical Forestry Action Programme. This programme, established in 1985 by the World Bank, FAO and the World Resources Institute, provides a mere framework for action, rather than a rigid organisation.[59]

Existing Institutions

(a) UN Institutions
One way in which intergovernmental organisations might be improved is, of course, in their collection and dissemination of data on global warming and the conduct of further scientific research. Thus, for example, Working Group I of the IPCC called for both the maintenance of the World Weather Watch Programme of the WMO and the approval of the establishment of a global ocean observing system. President Gorbachev has said that the Soviet Union is 'ready to co-operate in establishing an international space laboratory or manned orbital station designed exclusively for monitoring the state of the environment'.[60] The 'International Conference on Global Warming and Climate Change: Perspectives from Developing Countries', a conference designed to involve Third World experts in the debate, proposed the establishment of regional climate monitoring and management boards, consisting of the heads of equivalent national boards.[61]

Far more important, however, is the possible enhancement of the formal status and the roles of individual institutions. There would appear to be little need to alter the status of most of the specialised agencies, but the obvious candidate for an enhanced status, which would reflect the enhanced position of its governmental 'equivalents' in many countries, is UNEP.[62]

In May 1989, the UNEP Governing Council held a special session to discuss both the question of enhancing its status and the priority of its future activities.[63] The debate continues, but already a number of changes have been instituted to improve the catalytic and co-ordinating role of UNEP. The Bureau of the 15th Session of the Governing Council was mandated to meet the bureaux of the corresponding organs of the other specialised agencies in order to improve co-operation. In addition, the Environment Fund allocation was increased in 1989 to US$68 million, with a supplementary programme of US$32 million, US$4 million of which was earmarked for global warming purposes. It is aimed to increase the

Fund's regular level to US$100 million by 1991.[64] A number of countries, including the UK, have already increased their contributions to UNEP, and the UK has suggested that 'we may even need to consider in the longer term whether it might be promoted into a specialised agency'.[65]

Whether or not this occurs, the questions of changes to the system of vertical and horizontal accountability and co-operation arise. If environmental matters are to be given higher priority within the UN system, at the very least, it would seem desirable to rationalise the manner in which UNEP reports to the central organs. The UK has suggested that a 'Committee of the General Assembly might also be created which would have environmental matters as its prime responsibility'.[66] Indeed, rather than create an eighth committee, it would be sensible to transfer the work of one of the under-used ones, such as Committees Two or Four, to another one and thus to make that committee the 'Environment Committee'. UNEP should be required to report directly, and separately, to both UNGA and ECOSOC. Because reform of the Committees of UNGA can be effected without amendment to the UN Charter, this suggestion seems to hold some prospect for success.

Some hurdles would need to be cleared, however. The Non-Aligned Movement has different political priorities than the industrialised states and might even be said to regard proposed responses to global climate change as, in part, a Western device to keep them underdeveloped. Thus, any proposed reforms would not in practice be easily attainable.

A more radical suggestion is that of the USSR, which has, according to Fen Osler Hampson, adopted what he calls a Western idea

> to convert the [all-but defunct] Trusteeship Council [one of the UN's primary organs] into an Environmental Trusteeship Council composed of senior officials of member-countries on the Council who would review important environmental questions and co-ordinate the work of the various United Nations agencies dealing with environmental concerns.[67]

This is, it seems in addition to, and to be distinguished from, an earlier proposal made by the Soviet President in December 1988 'to set up a centre for emergency environmental assistance the function [of which] would be to send international groups of experts without delay to areas with a badly deteriorating environment'.[68] This proposal was later refined, in a letter addressed by the then Soviet Foreign Minister to the Secretary-General of the UN, into a specific proposal for a 'Council for Emergency Environmental Assistance', to be established within the UN Secretariat.[69] It was proposed as a response to the Chernobyl disaster – seeking a 'reaction' agency able to alleviate the chances and effects of sudden environmental disasters, rather than a putative institution that might be employed in the battle against more general and gradual negative environmental processes, such as global warming.

The Soviet proposal to reform the Trusteeship Council is perhaps the most surprising governmental proposal for reform of the UN system to date. Such a reform would certainly necessitate amendment of the UN Charter, a prospect which threatens to open a Pandora's box and which the Permanent Members of the Security Council, and in particular those whose status as such is most vulnerable – the UK and France – generally seek to avoid. In addition, it might be suggested that it would involve a degree of overkill, because it would appear to elevate environmental issues, whatever they are taken to be, to an institutional status rivalling fundamental issues of peace and security as well as questions of economics and development. It is perhaps significant that neither of these two Soviet ideas were pursued further in the former Soviet Foreign Minister's more recent statement on the subject of the environment before the UN.[70]

In addition, it would still be necessary for such matters to be discussed in UNGA and ECOSOC, in as far as the environment impinges upon issues within their remit. Thus, little would appear to be gained by the exercise, unless the Council was given some degree of enforcement powers in addition to those of discussing and encouraging initiatives. In any event, such a proposal is unlikely to command a sufficient degree of support for the same reasons as those discussed above in relation to UNGA reform.

The UK and USSR have also suggested additional involvement of the Security Council in environmental issues. The UK has suggested that it 'could take up environmental issues from time to time under Article 34 of the Charter'.[71] The former Soviet Foreign Minister has said:

> Since we are speaking of a major component of international security, I would add that political ecology requires the involvement of the Security Council in solving problems and activating such tools as transparency and strict international monitoring.[72]

The UK proposal would not seem to alter the Security Council's existing powers in relation to international peace and security, and the Soviet proposal's meaning is entirely unclear. It is highly unlikely that any proposal for reform involving amendment to the Charter will succeed, and the Council under an unamended Charter would seem to have no meaningful role in relation to environmental matters *per se*, as opposed to international disputes arising out of 'environmental' issues. The Security Council is an executive body and does not concern itself with the formulation of general strategies, as does UNGA. If it were to be given an environmental role, this would properly be limited to the relief and management of specific environmental disasters at most. It remains, however, that even this is far better taken care of in the context of a subsidiary programme of UNGA, rather along the lines of the United Nations Disaster Relief Organisation, than in the direct involvement of a primary organ of the UN.

Turning to horizontal co-operation among specialised agencies, it does not appear that any new mechanisms are necessary, and it is entirely correct that UNEP should continue to play the role of co-ordinator and provider of the Secretariats of DOEM and CIDIE. The participation of the World Bank in DOEM is essential, and what might follow upon UNEP's enhanced status (as a result of the conclusion of a global agreement on climate change) is its movement from a mere co-ordinating and catalytic role to a more directive one. It would also appear sensible to have UNEP representation on the board of a 'World Climate Fund',[73] if one is established, in conjunction with the World Bank and other agencies.

The World Bank already appears to have taken steps to reflect a greater awareness of the environmental aspects of development. It is difficult to contemplate any constitutional change to this and other existing international and regional financial institutions. They should concentrate on the continuation and improvement of 'green' policies in relation to the making of loans, the establishment of new 'green' funds and the enhancement of existing 'green' funds. They should participate in DOEM, as well as CIDIE. And they should co-operate with other agencies, such as FAO, in specific projects connected with climate change, such as the Tropical Forestry Action Plan.[74] Any World Climate Fund should be integrated into the World Bank's Global Environmental Facility, subject to safeguard against 'creeping conditionality'.

Nevertheless, it should be noted that there was strong pressure from environmentalist non-governmental organisations, before the founding of the new Bank for European Reconstruction and Development in 1990, to incorporate a particularly strong environmental mandate into its constitution and even to reflect the strength of environmental concerns about Eastern Europe in its name. It can be expected that such pressures will also be brought to bear upon such future institutions.

(b) Organisations of Industrialised Nations

It is difficult to envisage any useful reforms which might be made to the OECD, UNECE, EC or Group of Seven in view of the impending conference on global warming. It could be said, however, that these organisations are likely to have to show leadership during and after the negotiations in relation to both the enforcement of the regime and the control of bilateral and regional aid to ensure that development projects are 'green'.

(c) Other Intergovernmental Organisations

It is unclear what useful reforms might be made to other institutions, like the Non-Aligned Movement and the Commonwealth. Their role might be seen as both participants in the continued search for information about climate change and facilitators of information exchanges between the industrialised and developing countries about their concerns over the convention and its implementation.

CONCLUSION

The most likely institutional reform arising from the current developments concerning climate change, therefore, is the enhancement of the status and role of UNEP and the rationalisation of its place within the UN system. Because this will take place within the overall context of what is as much a legal as a policy-oriented negotiation, this is likely to be accompanied by an increase in the importance of international environmental law or, to be more precise, conventional international environmental law in the work of UNEP. The Climate Change Convention, if it is successful, will probably come to be regarded in UNEP as a possible model for further initiatives in the field.

Notes

The author wishes to acknowledge the research assistance of Lynne Jurgielewicz.

1. The USSR called for the holding of the International Conference on Environment and Development (*infra*) at the level of Heads of Government or State. See statement of Eduard Shevardnadze, Soviet Foreign Minister, before the UN General Assembly on 29 September 1989, UN Document A/44/PV.6, p. 46.

2. This is not meant, in any way, to underestimate the challenges of preserving the earth's ozone layer. As many commentators have noted, there may yet be significant challenges for the international community on this issue. However, because the ozone layer issue is much less complex than the global warming issue, coupled with the fact that an international legal regime is currently being developed to protect the ozone layer, a focus upon the global warming issue is justified. See also *op. cit.*, in notes 47, 48, 49 and accompanying text; and Peter M. Morrisette, 'The Evolution of Policy Responses to Stratospheric Ozone Depletion', *Natural Resources Journal* (Vol. 29, No. 3, Summer 1989) pp. 793–820.
3. World Commission on Environment and Development, *Our Common Future* (Oxford: Oxford University Press, 1987) paragraphs 31–2.
4. See, for example, P. Scharlin, *Reshaping Institutions to Meet Environmental Crises: Beyond Business as Usual*, Report of the Fifth Talloires Seminar on International Environmental Issues (Talloires, France: Tufts European Center, 14–18 May 1989) p. ii.
5. The original French draft envisaged a Declaration and 'ensuing legal provisions' acknowledging the commitments, *inter alia*, to

> accept in principle the establishment under the auspices of the United Nations of an authority which, in the context of the preservation of the earth's atmosphere shall be responsible for combating any further warming of the atmosphere . . . some decisions of which, regarding, *inter alia*, the determination of standards, may on occasion be adopted by a qualified majority; [and to accept] the principle of economic sanctions, inflicted subject to control exercised by the International Court of Justice on goods produced under conditions with negative impact on the atmosphere.

6. See UN Document A/44/340–E/1989/120, annex. The Declaration required Signatories to acknowledge and promote certain principles, including:

> [the] principle of developing, within the framework of the United Nations, new institutional authority, either by strengthening existing international institutions or by creating a new institution, which, in the context of the preservation of the Earth's atmosphere, shall be responsible for combating any further global warming . . . and shall involve such decision-making procedures as may be effective even if, on occasion, unanimous agreement has not been achieved [and] the principle of appropriate measures to promote the effective implementation of and compliance with the decisions of the new institutional authority, decisions which will be subject to control by

the International Court of Justice (ICJ).

The emphasis in relation to the role of the ICJ moved from the question of liability to the question of implementation of standards.

7. Statement of Sir Crispin Tickell before ECOSOC, 9 May 1989.
8. Lynton K. Caldwell, *International Environmental Policy: Emergence and Dimensions* (Durham, NC: Duke University Press, 1984) p. 63.
9. UNGA Resolution 2997 (XXVII) (15 December 1972), which established UNEP. See also UNEP Environmental Law Unit, *Environmental Law and the UN Environment Programme* (Nairobi: UNEP, 1985) p. 3.
10. Caldwell, *op. cit.*, in note 8, p. 63.
11. For ACC's involvement in climate change issues, see UNEP (GC.15/8/Add.3) paragraphs 14–22. For further information on CIDIE, see CIDIE Secretariat, *Action and Inter-action; the Role and Potential of CIDIE*, Volume 1 (Nairobi: CIDIE, 1989).
12. *The Guardian (London)*, 12 October 1990. The initial capital is likely to be 1 billion Special Drawing Rights.
13. UNEP has, for example, written the Environmental Operational Guidelines for the Policies and Procedures Manual of UNDP.
14. See, generally, U.P. Thomas, *The United Nations Environment Programme – Constraints and Strategy in the Context of 1992* (Washington, DC: Proceedings of the 31st Annual Meeting of the International Studies Association, 10–14 April 1990).
15. UNESCO, IOC, UNDP, WMO and others participate in the WCP, which was created by the Eighth World Meteorological Congress in 1979, following the First World Climate Conference. Operative paragraph 4 of UNGA Resolution 43/53 (27 January 1989) calls upon the relevant UN organs to give the WCP high priority. Sub-components include: the World Climate Impact Studies Programme (UNEP); the World Climate Research Programme (WMO-International Council of Scientific Unions [ICSU, a non-UN body]), which in turn has three sub-programmes and is linked to others; the Climate Computer (WMO); the World Climate Data Programme's Climate System Monitoring Programme (WMO, UNEP); and the Climate Change Detection Programme (WMO). Individual UN agencies also run independent programmes, such as the Global Information and Early Warning System on Food and Agriculture (FAO). These are, of course, complemented by programmes of non-governmental international scientific bodies, such as the ICSU, which involve co-operation with UN agencies, such as the International Geosphere-Biosphere Programme (IGBP), launched by ICSU in 1986.
16. This was launched in 1986 and involves ICSU co-operation with WMO, UNESCO and UNEP. Like the World Climate Research Programme (WCRP), it also possesses three sub-programmes.
17. See operative paragraph 10 of UNGA Resolution 43/51 (22 December

1989).

18. See the First Assessment Report and the Policy Makers' Summaries of the three working groups (June 1990). The working groups are concerned with: 'Scientific Assessment of Climate Change' (WG I), 'Potential Impacts of Climate Change' (WG II) and 'Formulation of Response Strategies' (WG III).
19. See operative paragraph 10 of UNGA Resolution 44/207 (22 December 1989), and UNGA Resolutions 43/196 (20 December 1988) and 44/228 (22 December 1989).
20. World Commission on Environment and Development, *op. cit.*, in note 3, paragraphs 105–6.
21. Reprinted as revised in *International Legal Materials* (Vol. 19, 1980) p. 524. It was concluded by the World Bank, five regional development banks, the EC Commission, the Organization of American States, UNEP and UNDP. The European Investment Bank signed in 1983.
22. See, for example, P. Muldoon, 'The International Law of Ecodevelopment: Emerging Norms for Development Assistance Agencies', *Texas International Law Journal*, (Vol. 22, 1987) p. 1.
23. *OECD and the Environment* (Paris: OECD, 1986) p. 6.
24. *UNECE Symposium on Problems Relating to the Environment: Proceedings and Documents of a Symposium Organised by the ECE, Prague, Czechoslovakia, 2–5 May 1971* (New York: ECE, 1971).
25. 'Bergen Conference: Ministerial Declaration on Sustainable Development', *Environmental Policy and Law* (Vol. 20, No. 3, 1990) p. 101.
26. Germany has set itself a target of reducing carbon dioxide emissions by more than 25 per cent from present levels by 2005 and Denmark of 20 per cent by the same date. *The Financial Times (London)*, 8 November 1990, p. 3. See also 'A Soft Target in the Greenhouse', *The Independent (London)*, 8 October 1990, p. 17.
27. Even the national policies adopted by some of these states might be internally incompatible. See, for example, 'Green to a fault', *The Economist* (16 June 1990) pp. 74–5, concerning Sweden.
28. See, for example, Council Directive 88/609 (24 December 1988) on the Limitation of Emissions of Certain Pollutants into the Air from Large Combustion Plants.
29. A Working Party consisting of representatives of the government departments responsible for taxes, finance and the environment was established in November 1989 and first met in February 1990. See *The Economist* (17–23 March 1990) p. 66.
30. EC Council Resolution on the Greenhouse Effect and the Community, 89/C 183/03 (21 June 1989).
31. It will be difficult, for example, to obtain Brazilian, Chinese and Indian participation in an effective regime without substantial transfers of resources or technology to them from developed countries. China plans to double its use of coal by the end of the century and India to triple its

use. Because this is likely to result in their output of carbon dioxide exceeding that presently produced by all OECD countries, this throws doubt upon the efficacy of an alternative OECD or regionally based 'mini regime'.

32. UN Document A/44/361, annex.
33. UN Document A/44/551–S/20870, annex.
34. UN Document A/44/673, annex.
35. *Op. cit.*, notes 13 and 14.
36. World Climate Programme Impact Studies, *Developing Policies for Responding to Climatic Change* (WMO/TD-No. 225) p. i.
37. See, for comment, D. Everest, *The Greenhouse Effect: Issues for Policy Makers* (London: Royal Institute of International Affairs and Policy Studies Institute, 1988) p. 9.
38. World Climate Programme Impact Studies, *op. cit.*, in note 36.
39. *Report of 17th Session of the General Assembly* (San Jose, Costa Rica: IUCN, 1–10 February 1988) paragraph 17.14.
40. *International Herald Tribune*, 27 July 1990, p. 20.
41. Svante Arrhenius, 'On the Influence of Carbonic Acid in the Air upon the Temperature on the Ground', *Philosophical Magazine* (Vol. 41, April 1896) pp. 237–76.
42. See also UNGA Resolution 42/184 (11 December 1987).
43. See WG I Policy Makers' Summary, *op. cit.*, in note 18.
44. About 300 scientists and policy-makers from 48 countries, UN organisations and NGOs attended. See *The Changing Atmosphere: Implications for Global Security* (Geneva: WMO, 1989).
45. *Protection of the Atmosphere* (Ottawa, Canada: Statement of the Meeting of Legal and Policy Experts, 22 February 1989) p. 1.
46. In particular, negotiations are continuing on the preservation of biological diversity, which are a major distraction from global warming issues. See, for example, Petsouk 'The Role of UNEP in the Development of International Environmental Law', *American University Journal of International Law and Policy* (Vol. 5, 1990) pp. 386–8.
47. *International Legal Materials* (Vol. 26, 1987) p. 1541.
48. The Montreal Protocol on Substances that Deplete the Ozone Layer, *ibid.* See also the Helsinki Declaration on the Protection of the Ozone Layer (UNEP/OzL. Pro. 1/5, 2 May 1989), appendix I; and the statement of the London Meeting to review the Protocol, 19–29 June 1990.
49. The London Meeting created a Global Fund to be managed by a committee of 14, including representatives from UNEP, UNDP and IBRD. This fund, totalling US$160 million, will rise to US$240 million in the event that China and India accede to the Protocol. These two countries are taking steps to do so.
50. *Report of Special Session of the Governing Council of UNEP* (GC.15.12, 4 July 1989).
51. WG III's Policy Makers' Summary, *op. cit.*, in note 18, p. 3.

52. For example, UNGA Resolution 43/53, *op. cit.*, in note 15; and UNGA Resolution 44/207, *op. cit.*, in note 19.
53. *Ibid.*, p. v.
54. These proposals for low standards related to the establishment of a carbon tax or insurance fund. It is interesting to note that West Germany, normally expected to be a leader in favour of high standards, lined up behind the disappointing proposals. See, for example, Stephen H. Schneider, 'The Costs of Cutting – or not Cutting – Greenhouse-gas Emissions', in Jeremy Leggett (ed.), *Global Warming: The Greenpeace Report* (Oxford: Oxford University Press, 1990) p. 186.
55. UN Document Report A/44/48 (20 March 1990).
56. Statement of Tickell, *op. cit.*, in note 7, p. 2.
57. See, for example, notes by Brown-Weiss, D'Amato and Grundling in *American Journal of International Law* (Vol. 84, 1990) pp. 190–212; Edith Brown-Weiss, *In Fairness to Future Generations* (New York, 1988); and Edith Brown-Weiss, 'The Planetary Trust: Conservation and Inter-generational Equity' *Ecology Law Quarterly* (Vol. 11, 1984) p. 495.
58. See, for example, Brown-Weiss, 'The Planetary Trust', *op. cit.*, in note 57, p. 564.
59. *The Financial Times (London)*, 14 June 1990.
60. Statement of M. Gorbachev before UNGA on 7 December 1988, UN Document A/43/PV.72, p. 19.
61. *Statement of the Conference* (New Delhi, India: 21–23 February 1989) paragraph 7. The conference was organised by the Indian Tata Energy Research Institute, The Woods Hole Research Centre, UNEP and the World Resources Institute.
62. In the US, the post of Administrator of the Environmental Protection Agency is being considered for elevation to Cabinet rank. In Brazil, a prominent environmentalist, Jose Lutzemberger, has been conspicuously appointed as Minister for the Environment. In the UK, environmental issues have taken on an enhanced status within the government, as illustrated not least by the seminar on climate change organised by the Prime Minister for her Cabinet in April 1989. In addition, the Foreign and Commonwealth Office has reorganised its departmental structure to reflect the growing importance of environmental issues.
63. Decisions of UNEP Governing Council Nos. 15/1 and 15/4 (25 and 26 May 1989); Report of Council, 15 S., GAOR 44 S., Supp. No. 25 (A/44/25).
64. *Ibid.*
65. Statement of Tickell, *op. cit.*, in note 7, p. 2.
66. *Ibid.*
67. Fen Osler Hampson, 'Climate Change: Building International Coalitions of the Like-Minded', *International Journal* (Vol. 45, No. 1, Winter 1989–90) p. 70.

68. Statement of Gorbachev before UNGA, *op. cit.*, in note 60.
69. Letter dated 30 April 1989, Annex to UN Document A/44/264 E/1989/73 (2 May 1989). The elaborated purpose of the proposed Centre, which would be headed by an Under-Secretary-General, would be 'to organise international co-operation in critical environmental situations'. The experts would examine the situation on site and draw up recommendations on the limitation and elimination of the consequences of the disaster in question. Their assessments would be widely publicised. As the Centre developed, it would also provide necessary technical assistance to the states concerned. The feature of the proposal that is probably of most relevance to the debate on the institutional consequences of global warming is the suggestion that the groups of experts 'would include lawyers to consider legal problems that may arise as a result of environmental disasters with transboundary effects'.
70. Statement of Shevardnadze, *op. cit.*, in note 1, p. 46.
71. Statement of Tickell, *op. cit.*, in note 7, p. 2.
72. Statement of Shevardnadze, *op. cit.*, in note 1.
73. See World Commission on Environment and Development, *op. cit.*, in note 3, paragraph 107, which talks of a 'World Conservation Bank'.
74. M. Bowman, *Legal Implications of Global Climate Change: Tropical Forests* (paper for the Environmental Section Meetings on the Legal Aspects of Global Warming of the British Branch of the International Law Association, 1990).

8 North-South Issues, the Common Heritage of Mankind and Global Environmental Change

Kilaparti Ramakrishna

NORTH-SOUTH ISSUES

Gro Harlem Brundtland, Prime Minister of Norway and Chairperson of the World Commission on Environment and Development, was awarded the Indira Gandhi Prize for Peace, Disarmament and Development in New Delhi on 20 September 1990.[1] This event is significant, because Indira Gandhi, the then Prime Minister of India, led the Indian delegation to the first United Nations Conference on the Human Environment, held in Stockholm in 1972, and articulated what soon became the acknowledged position of the developing countries on environmental issues.

Indira Gandhi's position at the Stockholm Conference was that the developing countries are as much concerned about the urgent need to protect the environment as any other state, if not more, but that when issues of global environmental protection are competing for scarce resources needed to address the very means of improving living conditions, of providing food, water, sanitation and shelter, the developing countries are hard pressed to pay much attention to environmental issues. She asked unequivocally: 'How can we speak to those who live in villages and in slums about keeping the oceans, rivers and the air clean when their own lives are contaminated at the source?'[2]

Despite this dilemma, the developing countries, by and large, have since enacted environmental legislations, established administrative structures for environmental preservation and, in some cases, even gone to the extent of providing constitutional guarantees to protect the environment for present and future generations. But these actions have often addressed environmental problems of immediate concern, rather than those of a longer time horizon. Implicit in many of these actions is the acknowledgment that the industrialised countries are primarily responsible for many global environmental problems. Decision-makers in developing countries believe that if action by all nations is warranted to deal effectively with global environmental problems, then the industrialised countries will need to make every effort to ensure that such action will not place additional monetary burdens on developing countries.[3]

Many concerns voiced by Indira Gandhi, and subsequently by the leadership in several developing countries, were articulated very effectively by Brundtland and her Commission members when they produced the much acclaimed report entitled *Our Common Future*. At the Toronto Conference on 'The Changing Atmosphere: Implications for Global Security' in June 1988 – the first large-scale discussion of policy options for climate change – Brundtland drew from this Report in much of her keynote address. She pointed out that there is 'a need for a fresh impetus in international cooperation. Development aid and lending must be increased, and the debt crises must be resolved.' She added that 'the ultimate goal must be to forge an economic partnership based on equitable trade and to achieve a new era of growth, one which enhances the resource base rather than degrades it'. She concluded by stating that 'the mission must be to make nations return to negotiations on the global issues after years of decline in real multilateralism'.[4]

Awarding the Indira Gandhi Prize to Brundtland, the President of India pointed out that

> if we are to effectively deal with global environmental problems, we have to forge a partnership which seeks simultaneously to protect and serve the environment taking care, at the same time, of the special requirements of developing countries . . . The environment must not be dealt with as yet another North-South issue.[5]

At the same occasion, the Indian Prime Minister declared that: 'Rich and poor, we must all work with each other to preserve the global environment. Our very survival is at stake.'[6] It is perhaps sentiments of this nature that effectively warded off possible large-scale criticism and rhetoric from either the Third World participants or the nongovernmental organisations taking part in the ongoing discussions towards adopting appropriate policies to deal with global climate change. The Toronto conference document referred to above notes:

> The countries of the industrially developed world are the main source of greenhouse gases and therefore bear the main responsibility to the world community for ensuring that measures are implemented to address

> the issues posed by climate change. At the same time, they must see that the developing nations of the world, whose problems are greatly aggravated by population growth, are assisted and not inhibited in improving their economies and the living conditions of their citizens. This will necessitate a wide range of measures, including significant additional energy use in those countries and compensating reductions in industrialized countries. The transition to a sustainable future will require investments in energy efficiency and non-fossil energy sources. In order to ensure that these investments occur, the global community must not only halt the current net transfer of resources from developing countries, but actually reverse it. This reversal should embrace the relevant technologies involved, taking into account the implications for industry.[7]

Mostafa Tolba, Executive-Director of the United Nations Environment Programme, joined Brundtland when he pointed out that global warming and climate change present a window of opportunity for both the industrialised and the developing countries to address a variety of priority items, including trade imbalances, debt relief, technology transfer, technical assistance and financial assistance.[8]

From a practical and legal standpoint, however, there is a need to ascertain how these sentiments might be acted upon, particularly in terms of the commitments to be made by the industrialised world and the proposed international agreements to be adopted by the world community. Despite an occasional comment in newspapers to the contrary, one sees a tremendous surge of interest among the developing countries to participate, in a variety of ways, in the various international and intergovernmental conferences that address the issue of climate change. Developing countries have been particularly active in the activities of the Intergovernmental Panel on Climate Change (IPCC) since its inception in November 1988. Its Fourth Plenary in Sundsvall, Sweden in 1990 was attended by 72 countries, many of them developing countries.

Historically, developing countries were reluctant to subscribe to an international agreement that sought to codify the principles of state responsibility and liability, which are considered essential to the effective implementation of any strategy to reduce pollutants globally. Even at the 1972 Stockholm Conference, the principal reason attributed to developing country participation was that they could learn how to promote developmental activities in their countries while, at the same time, addressing environmental concerns.[9]

There is a view held in several quarters that the United Nations

Convention on the Law of the Sea – an international agreement which places great emphasis on environmental protection, principles of state responsibility and liability, and is concerned with the special circumstances of the developing countries – should be used as a basic framework for a convention on the atmosphere. Therefore, an evaluation of the approach adopted by the developing countries during the negotiations for that treaty can yield insights which may prove relevant to the climate change debate. Likewise, several scholars maintain that the Vienna Convention and the Montreal Protocol for the protection of the ozone layer are useful models to emulate. While the view that the Law of the Sea or the protection of the ozone layer experiences should be repeated for a law of the atmosphere has some obvious merits, it will be concluded that such a suggestion is not without its drawbacks. This point is made with special reference to the principle of the Common Heritage of Mankind.

THE LAW OF THE SEA CONVENTION

The international environmental movement that originated with the preparations leading to the 1972 United Nations Conference on the Human Environment helped to sustain the activities of national environmental organisations. These groups, in turn, were able to ensure that environmental concerns remained on national agendas.[10] To move from the general to the specific, the Law of the Sea Convention will now be studied in two manners. First, the special provisions incorporated to enlist the support of the developing countries will be described, and second, it will be considered whether this convention, with some modifications, might be able to address the increases in the concentration of atmospheric trace gases.

Part XII of the Law of the Sea Convention of 1982 contains provisions for the protection and preservation of the marine environment. Included in this section are measures to prevent the pollution of the oceans by pollutants carried through the atmosphere.[11] In this regard, the Law of the Sea Convention provides a form of atmospheric pollution control. The question arises as to whether or not this framework for the protection of the atmosphere can be used to control the emissions of greenhouse gases and thus prevent environmental damage.

The drafting of Part XII of the Convention was largely the work of Committee III. Compared to Committees I and II, Committee III's work in general, and on matters related to the protection of the marine environment in particular, produced very little debate.[12] Committee III was the first to report that 'substantive negotiations on all matters assigned to the Committee had been completed'.[13] It also declared that it 'attained a level of agreement which can be considered to offer a substantially improved prospectus for consensus'.[14] A review of the relevant literature also demonstrates that the protection and preservation of the marine environment was of peripheral interest to the negotiating states.[15] Provisions regulating the potential mining of the deep sea-bed dominated the discussion, while issues such as the breadth of the territorial sea, exclusive economic zone (EEZ) regulations and fishing rights also provoked debate.

One explanation for the absence of any protracted debate over the protection of the marine environment is that Part XII represents a codification of accepted customary international law on the subject.[16] Another possible explanation is that the provisions of Part XII (Articles 192–237) contain important qualifications, which lessen the responsibilities of states with regard to pollution control. According to Article 194(1) of the Convention:

> States shall take, individually or jointly as appropriate, all measures consistent with this Convention that are necessary to prevent, reduce and control pollution of the marine environment from any source, using for this purpose the best practicable means at their disposal and in accordance with their capabilities, and they shall endeavour to harmonize their policies in this connection.[17]

Similar provisions, regarding the control of pollution from land-based sources, are included in Article 207. Article 207(4) reads:

> States, acting especially through competent international organizations or diplomatic conference, shall endeavor to establish global and regional rules, standards and recommended practices and procedures to prevent, reduce and control pollution of the marine environment from land-based sources, taking into account characteristic regional features, the economic capacity of developing States and their need for economic development. Such rules, standards and recommended practices and procedures shall be re-examined from time to time as necessary.[18]

These qualifications were incorporated on the demand of the developing countries, because they strongly felt that the obligation to implement international standards would place a 'disproportionately heavy burden' on them.[19] These qualifiers in essence mean that the developed and developing countries are treated differently in the interpretation and application of some of the provisions. This is often referred to as the 'double standard' clause of the Convention. It must be pointed out here that pollution, whether emanating from the developed or developing countries, can cause equal harm to the marine environment. At the same time, the developing countries, in their present stage of economic development, cannot be expected to bear the additional strain of keeping the marine environment free from pollution. The more appropriate action would have been to make provision for meeting the additional expenditure, rather than to grant concessions in standard setting.

Assuming that the international community is successful in establishing environmental standards with regard to the marine environment, there will remain the difficult task of achieving comprehensive and effective enforcement. According to Articles 213 and 222, states will be responsible for the enforcement of all rules and regulations for the protection and preservation of the marine environment from pollution originating from atmospheric and land-based sources.[20] States, therefore, will be the sole authority with the power to enforce environmental standards within their jurisdiction. The actual enforcement of established rules and regulations will thus depend on the ability, resource allocation and political will of each state.[21] In practice, this could lead to widely varied enforcement of established standards for pollution control.

The possible consequences of the inclusion of these provisions in Part XII of the Convention has led some to doubt their efficacy as a means of protecting the marine environment:

> Several factors render [Part XII] useless as a pollution control device. First, and most obviously, the text lacks any form of mechanism either to assure that international standards are established or enforced. Second, the text demands that in establishing international standards, international organizations take into account 'the economic capacity of developing countries and their need for economic development'. Finally, the extension of the economic zone, would make international controls over land based sources even more difficult . . . There is little possibility that most pollution from land-based sources could be traced to its original source. Therefore, a coastal state could not be held liable for damages created by its failure to control land-based sources.[22]

These weaknesses render the Law of the Sea Treaty ineffective as an example of an instrument that might be developed to protect the atmosphere from increased emissions of greenhouse gases.

Before leaving the Law of the Sea Convention, it is important to note some significant factors that may be of value to the future development of a regime for the protection of the earth's atmosphere. First, as previously mentioned, the Convention represents a codification of customary international law. It has, therefore, contributed to the progressive development of environmental law and has become a potential catalyst for the continued evolution of international environmental law, leading to stricter and more specific pollution regulation and enforcement. Second, Part XII achieved a consensus among the nations of the world, including the United States, which had not been previously attained in this field.[23] In so doing, the provisions of Part XII represent an acknowledgement by the world community that the development of international law to protect the global environment is necessary. Third, Part XII is comprehensive in respect to the inclusion of states: 'It is quite clear that the provisions of [Part XII] are meant to be universal, applicable to all States as generally accepted customary international law.'[24] Finally, the Convention is capable of adapting to changing environmental conditions and standards:

> The environmental provisions of the Law of the Sea Convention are dynamic in character: they build on the accomplishments of the past . . . they add to the structure where there were some gaps, and they establish a method for incorporating future developments into the Convention without the necessity of constant amendments.[25]

Efforts to protect the global environment from the consequences of greenhouse gas emissions will benefit from the momentum generated by the Law of the Sea negotiations. The universality and dynamic character of Part XII should be included in the establishment of a regime to protect the earth's atmosphere. While Part XII is not a viable solution to the problem of greenhouse gases, it offers a precedent upon which the world community can build to establish effective, strict and universal protection of the atmosphere.

OZONE LAYER PROTECTION

As with the Law of the Sea Convention, the Vienna Convention for the Protection of Ozone Layer[26] and the Montreal Protocol on Substances that Deplete the Ozone Layer[27] will be studied in two ways. First, the special provisions incorporated to enlist the support of the developing countries will be described; and second, it will be considered whether this convention, with some modifications, might be able to address the increases in the concentration of atmospheric trace gases.

A reading of the preamble of the Vienna Convention provides the circumstances under which the agreement was adopted. There was sufficient reason to believe that damaging the ozone layer would have potentially harmful impacts on human health and the environment. The Parties to the Vienna Convention were equally convinced that there was a need for obtaining 'proof' by continued research and systemic observations before specific commitments to reduce ozone-depleting substances, chlorofluorocarbons (CFCs), were agreed upon. This response was largely dictated by the perceived impact on the CFC industry as well as the particular needs of the developing countries. However, the scientific proof was no more compelling a little over two years later, when the Montreal Protocol was signed imposing 'substantial short-term economic costs'.[28] While the Vienna Convention aims to promote exchanges of information, research and data on monitoring with the goal of protecting human health and the environment against activities that have an adverse effect on the ozone layer, the Montreal Protocol calls for significant reductions in the use of chloroflurocarbons and halons.

This feature alone distinguishes the Vienna Convention and the Montreal Protocol from other environmental agreements, in that these two were concluded before any directly attributable damage was suffered by humankind as a whole or by any part of it due to the ozone 'hole'. In this sense, the world community for the first time adopted preventive and precautionary measures to address environmental dangers predicted by scientists.

The special circumstances of the developing countries were acknowledged in the Montreal Protocol in the following terms. Developing countries within a certain per capita consumption limit may delay their compliance with the reduction schedules outlined in the Protocol by ten years, while not exceeding a consumption level of 0.3 kg per capita. The industrialised countries agree to facilitate the developing countries' access to alternative substances or technologies and to financial aid in this regard. Like the Vienna Convention, the Montreal Protocol also recognises the developing countries' interests in the fields of research, development and exchange of information.

While these concessions were enough to attract some developing countries as signatories, India and China, to name but two countries among the developing world, called it iniquitous and have not yet joined the agreements. Even among those countries that have signed on as parties to the agreements, a feeling that they were 'caught napping' before and during the negotiations is gaining ground. If another opportunity were to present itself, many believe that they should react differently.

The obvious similarities between concerns about the damage to the ozone layer and the accumulation of the greenhouse gases may lead several to argue that the proposed negotiations on a climate change convention should be structured along the lines of the Vienna Convention. That is, they should begin with an umbrella convention followed by a set of protocols as nations become prepared to commit themselves to specific obligations. If the question is a comparison between the Law of the Sea Convention and the Ozone Layer Convention, one may conclude that the latter is more appropriate. But, when taken independently, the experience gained in the Vienna Convention appears to present as many new problems as any other convention in directly using it to address the climate change issue.

When one looks at the entrenched industrial interests, the similarities between the climate change and ozone layer issues become obvious. In the climate change issue, the position of these industrial interests is that agreements should wait until all the scientific uncertainties surrounding the issue are resolved. And also, they argue that action now would be costly, it would seriously damage lifestyles and stunt industrial growth. It is interesting to note that the very same set of excuses were used in the ozone layer negotiations to delay hard decisions. When nations went ahead with the Vienna Convention, however, the CFC industry learned very quickly to come up with substitutes and even announced a phase-out schedule ahead of governmental regulations in some countries.

In approaching the ozone layer depletion issue, it was clear to all governments involved that it was a problem caused by industrialised countries, because most CFC plants are located within their jurisdictions. It was also clear to industrialised nations that efforts should be taken to meet the CFC-related needs of developing countries. Sufficient incentives could be devised and coupled with the development of new technology to avert further damage to the ozone layer. Despite this, an effort was made to seek developing country participation in the negotiations. These negotiations, however, were not seen to be contingent upon achieving specific obligations from developing countries. When one looks at the complex set of issues to be addressed in the climate change negotiations and the important role that the developing countries will have to play to achieve any meaningful reductions in the greenhouse gas accumulations, it becomes obvious that the developing countries' participation is not only necessary to the successful conclusion of the climate change issue, but also integral to it.

THE PRINCIPLE OF THE COMMON HERITAGE OF MANKIND

The international community has established regimes for the protection and management of the environment of areas outside national jurisdictions. Recent examples of such regimes include the Agreement Governing the Activities of States on the Moon and Other Celestial Bodies, 1979, and the aforementioned United Nations Convention on the Law of the Sea, 1982. In both of these treaties, the 'Common Heritage of Mankind' (CHM) principle was incorporated to assert international control over regions beyond national jurisdictions. The central question addressed here is: What are the implications of the CHM principle for the issues of global environmental change, such as the control and reduction of greenhouse gas emissions?[29] To answer this question, the development of the CHM principle, its acceptance or rejection by various groups of states and its efficacy as an approach to areas beyond national jurisdictions will be discussed.

The concept of a Common Heritage of Mankind was voiced by the Maltese Ambassador, Arvid Pardo, before the United Nations General Assembly (UNGA) in 1967. 'Pardo told the world of the vast resource potential that lay beneath the oceans beyond the limits of national jurisdiction, and warned that its existence created a "basic political problem" that could no longer be avoided.'[30] The problem was the recognition that those resources constituted a Common Heritage of Mankind in which all states were entitled to share. Particularly to be avoided, in Pardo's view, was the possibility that technological capacity would determine which countries would acquire the oceans' wealth. Under such a formulation, it was said, the rich nations would get richer and the poor nations would become relatively poorer. Pardo proposed the establishment of an international regime: 'Our long term objective is the creation of a special agency with adequate powers to administer, in the interests of mankind, the oceans and the ocean floor beyond national jurisdiction.'[31]

The establishment of this 'special agency' was more a means to an objective than an objective in itself. The desired result was a safer and more peaceful world order and a more equitable international economic order.

> It is quite clear that, by making this [CHM] proposal, Malta had two purposes, one military, one economic. Originally the military object may have been uppermost in Ambassador Pardo's mind, but in recent years more attention has been paid to the economic aspect of his proposal. As he himself put it a few months after introducing the proposal:

> 'The resources of the seabed and of the ocean floor are as great as the resources on dry land. The seabed and the ocean floor are also of vital and increasing strategic importance. Present or clearly foreseeable technology permits the effective exploitation of virtually all of this area for military or economic purposes. Some countries may therefore be tempted to use their technical competence to achieve near unbreakable world dominance through predominant control over the seabed and the ocean floor. This, even more than the search for wealth, will impel countries with the requisite technical competence competitively to extend their jurisdiction over selected areas of the ocean floor. The process has already started and may lead to a competitive scramble for sovereign rights over the land underlying the world's seas and oceans, surpassing in magnitude and in its implications last century's colonial scramble for territory in Asia and Africa. The consequences will be very grave – at the very least a dramatic escalation of the arms race and sharply increasing world tensions, caused also by the intolerable injustice that would reserve the plurality of the world's resources for the exclusive benefit of less than a handful of nations. The strong would get stronger, the rich richer, and among the rich themselves there would arise: an increasing and insuperable differentiation between two or three and the remainder. Between the very few dominant powers, suspicions and tensions would reach unprecedented levels. Traditional activities on the high seas would be curtailed and, at the same, the world would face the growing danger of permanent damage to the marine environment through radioactive and other pollution. This is a virtually inevitable consequence of the present situation.'[32]

It is interesting to note that environmental protection was very much a peripheral issue in the formulation of the common heritage concept.

This sentiment, as adopted by the developing world and carried forth by the Group of 77, was reiterated in UNGA Resolution 2749, Declaration of Principles Governing the Sea-Bed and Ocean Floor, and the Subsoil Thereof, Beyond the Limits of National Jurisdiction, on 17 December 1970. This resolution

> solemnly declares that:
>
> 1. The sea-bed and ocean floor, and the subsoil thereof, beyond the limits of national jurisdiction (hereinafter referred to as the area), as well as the resources of the area, are the common heritage of mankind. . . .
>
> 4. All activities regarding the exploration and exploitation of the

> resources of the area and other related activities shall be governed by the international regime to be established. . . .
>
> 7. The exploration of the area and the exploitation of its resources shall be carried out for the benefit of mankind as a whole, irrespective of the geographical location of States, whether land-locked or coastal, and taking into particular consideration the interests and needs of developing countries.[33]

The concepts of international management of the resources of common areas, sharing of the benefits derived from the exploitation of such resources and the requirement to grant special consideration to the needs of the developing countries have become the foundation of the common heritage principle.

The main characteristics of the CHM principle include the following four basic elements: (1) the area under consideration cannot be subject to appropriation or national sovereignty; (2) all countries must share in the management of the area; (3) benefits reaped from the exploitation of the area's resources must be actively shared and distributed for the benefit of mankind; and (4) the use of the area must be reserved exclusively for peaceful purposes.[34] A possible fifth basic element is free and open scientific research in the area for all interested parties.[35]

As previously mentioned, the CHM principle has been included in the development of international law regarding mankind's exploration and possible exploitation of outer space. Article I of the Treaty on Principles and Use of Outer Space, Including the Moon and Other Celestial Bodies, 1967, states:

> The exploration and use of outer space, including the moon and other celestial bodies, shall be carried out for the benefit and in the interests of all countries, irrespective of their degree of economic or scientific development, and shall be the province of all mankind.[36]

The exact meaning of this paragraph was of concern to the US Government.

> As a result of the concern [regarding this paragraph] the United States Senate attached its understanding of its meaning at the time it gave its advice and consent respecting ratification. This understanding reads: 'It is the understanding of the Committee on Foreign Relations that nothing in Article I, paragraph I of the treaty diminishes or alters the right of the United States to determine how it shares the benefits and results of its

> space activities.' Following the entry into force of the Principles Treaty in 1967 down to the present, States have not asserted claims to specific acquisitions resulting from the space activities of other States. . . . On the other hand, space resource States have voluntarily shared or made available at nominal costs the materials which have been acquired.[37]

Thus, the states were to determine what and how they would share the benefits obtained from space exploration and exploitation. This issue was again addressed in the Agreement Governing the Activities of States on the Moon and Other Celestial Bodies, 1979, Article XI.[38] The specific organisational structure and regulations guiding the exploitation of the resources of these areas, however, have yet to be formulated. This is due to the fact that mankind has not yet acquired the technological capacity to harvest these resources on a commercially viable scale. The CHM principle, nevertheless, has played a small role in regulating the uses of space.

> Determination of the degree to which the benefits of the use of space and benefits derived therefrom must be shared will necessarily require the actuality of such activities in order for application. Nevertheless, the common heritage doctrine requires that information concerning scientific advancements and the products or services derived as a result of some particular use of space be made available to the global community on a non-discriminatory basis.[39]

Because the need has not yet arisen, the international community has not developed a mechanism to control these resources. The CHM principle with regard to outer space and celestial bodies, therefore, offers little guidance for the principle's application to the climate change issue.

The situation regarding the Law of the Sea, however, is vastly different. Because of the currently available and potentially exploitable resources of the oceans, the Law of the Sea has become the testing ground for the CHM principle.[40] The potential mining of these resources necessitated a clarification of the rights and obligations of states and private enterprises within the 'area' (the sea-bed and ocean floor beyond the limits of national jurisdiction). To this end, various articles within the United Nations Convention on the Law of the Sea establish an authority to manage deep sea-bed mining. Article 136 states that 'the Area and its resources are the common heritage of mankind'.[41] Article 137 establishes the legal status of the area both by denying states the right to claim sovereignty over any part of the area or its natural resources and by reserving these resources for the benefit of mankind as a whole.[42] The economic aspirations of the developing countries, as articulated by Pardo's 1967 address and the 1970 UNGA Resolution 2749, are included in Article 140(1):

> Activities in the Area shall be carried out for the benefit of mankind as a whole, irrespective of the geographical location of States, whether coastal or land-locked, and taking into particular consideration the interests and needs of the developing countries and peoples who have not attained full independence or other self-governing status as specifically provided for in this part.[43]

Finally, and most importantly, the Convention provides for the management of the resources of the area:

> [The] resource development of the deep sea-bed is to proceed under the aegis of an international institution known as the International Deep Sea-Bed Authority (Arts. 156–185), which will operate both as a global regulatory agency and as a deep sea-bed miner.[44]

Because the Law of the Sea negotiations represent the first real attempt to apply the common heritage principle to an international treaty in more than merely a philosophical manner, it offers an excellent opportunity to determine its usefulness, with regard to the atmosphere, for addressing the global climate change problem.

The CHM principle and its application to the resources of the area became the primary obstacle to a final consensus on the Law of the Sea Treaty. The controversy surrounding this principle is actually only symptomatic of a much larger struggle.

> This international conflict focusing on the law of the sea is only one dimension of a much wider confrontation between the affluent industrialised nations of the 'North' and the less developed, destitute countries of the 'South'. The latter demand a radical restructuring of the economic power relationship between the two groups of states in the name of the 'New International Economic Order' (NIEO).[45]

The potential economic implications of deep sea-bed mining, on such things as trading patterns, relative economic wealth and economic dependency versus development, forced the issue of international economic relations to the forefront of the negotiations. The content of these negotiations contained legal, economic and ideological components revolving around access to deep sea-bed exploitation. Included were questions of freedom of the seas, the status of the common area's resources (*res nullius* versus *res communis*) and methods of economic management (economic liberalism versus centralised management).

> The economic problem emerges from the differing philosophies of the developed and developing countries concerning exploitation of natural resources in common spaces. The United States, representative of the developed countries' viewpoint, prefers maximum freedom to exploit resources in these areas and to earn maximum revenues for itself and its nationals. . . . By contrast, the CHM concept, as interpreted by the Third World, stands for the notion that development of resources 'to which no single decision-making unit holds exclusive title' must be undergirded by safeguards and strict controls to ensure that the most needy countries benefit the most. Under this approach, the international equivalent of a government monopoly would carefully manage and regulate natural resources. This 'monopolistic' form of socialism would restrict U.S. economic activity in Antarctica [or the deep sea-bed and outer space], and would consequently represent an unprecedented move towards central management of the international economy.[46]

The crux of the concern over the application of the CHM principle to the deep sea-bed has been its effect on access to sea-bed minerals. The developed nations historically regarded

> the deep sea-bed resources to be *res nullius*, and therefore available, by virtue of freedom of the seas, for appropriation by any nation capable of exploiting them. The developing countries regard the sea-bed resources as *res communis*, a common heritage of mankind, to be shared by all irrespective of their technological capabilities.[47]

The respective philosophies of these two groups are articulated by the following statements of US Senator Patrick Moynihan and Mr. Joseph S. Warioba, Tanzanian delegate to UNCLOS III. As Moynihan argued:

> Do the developing nations understand that by entering into the [Law of the Sea] negotiations, and remaining faithful with them, the United States and the western nations generally have agreed to negotiate for, and in the bargaining sense, to pay for rights which exist in the absence of a treaty? . . . We did not have to do this. We have under existing international law . . . the perfect right to extract mineral resources from

> the deep ocean beyond the Continental Shelf. Moreover, we have the technology to do so.[48]

Warioba's statement underscores the developing world's frustration with the established international legal order as referred to by Moynihan:

> Freedom of the seas has ceased to serve the interests of international justice. It had become a catch word and an excuse for a few countries to exploit ruthlessly the resources of the sea, to terrorize the world and to destroy the marine environment. That type of freedom belonged to the old order and had outlived its time.[49]

The desire of the newly formed developing nations to change this aspect of the international legal order is derived from the perceived historical connotations of the 'freedom of the seas'. As Hasjim Djalal reasoned, freedom of the seas resulted in

> the conquest of what have now become independent developing countries was to a large extent facilitated by the principles of the freedom of the sea as practised at that time. The freedom of the sea in the past, in effect, meant the freedom of those who had the navy to control the oceans, thus controlling the world.[50]

These arguments outline the positions taken by the developing and developed states during the Law of the Sea negotiations. As the final text of the treaty shows, the developing states' position prevailed.

> [The] clearest impact of the NIEO upon the Convention rules governing the Area is in the complex provisions dealing with quantitative limits on the production of sea-bed minerals. These limits are designed to protect developing land-based producer countries against overabundant world mineral supplies which would reduce prices and result in lower export earnings.[51]

In addition to these production limits, the International Seabed Authority (ISA) would attempt to support commodity prices.

> The ISA's goal through such participation would be to stabilise world markets for raw materials and commodities in favor of the interests of the developing countries which are producing them. In particular, there has been controversy as to whether the ISA may become a party

> to the UNCTAD-conceived Integrated Program of Commodities (IPC) with respect to copper and manganese. As viewed by the ideology of economic liberalism, agreements of this kind are 'cartelization' and disregard consumers' interests.[52]

Inclusion of this form of the CHM principle in the Convention, however, has not put an end to the debate.

> For the doctrine of the CHM to be regarded in fact as an accepted principle of contemporary international law, at least three qualifying preconditions must be fulfilled. First, the legal content of the CHM must be so distinct and well-defined that the concept can be fully integrated into the corpus of international law. Second, resultant State practice must comply with the development of the CHM notion and, additionally, evidence of *opinio juris* must be demonstrated and evident. Third, the customary acceptance of the CHM as determined by State conduct and behavior must be manifest, or at least sufficiently broad-based to attest to the CHM's wide-spread acceptance. . . . The CHM today is neither the product of 'instant custom' nor *jus cogens*. Rather it is merely a philosophical notion with the potential to emerge and crystallize as a legal norm. The prime determinant of how, when or whether that norm-creating process will happen rests in the conduct of States. When sufficient State practice indicates clear, wide-spread acceptance of the CHM, its legitimacy in international law will be fixed.[53]

Still to be determined is the efficacy of the CHM principle as an approach to the global climate change issue. While it is difficult to determine the exact role the principle would play, it would seem to differ from its function in either the Law of the Sea Treaty or the law of outer-space experience. The benefits to be accrued in prior arrangements would be replaced by sacrifices incurred. Any international managing authority that might be created would have to oversee the allocation of responsibilities regarding the reduction of greenhouse gas emissions rather than the right to exploit resources. From the perspective of the developing countries, the positive aspect of benefit sharing would be replaced by the negative notion of burden sharing. Likely similarities with other applications of the CHM principle include the complicated but clear relationship between resource management and environmental concerns, and economic and social development.

An attempt to apply the CHM principle to a law of the atmosphere would encounter serious difficulties. To maintain the integrity of the four essential elements as described above,[54] vast changes in present national as well as international law would be required. The first element states that the region not be subject to appropriation or national sovereignty. The airspace and atmosphere within national boundaries, however, is presently included as a part of national sovereignty. The fourth element states that the area (that is, the atmosphere) be used exclusively for peaceful purposes. This, for obvious reasons, would create insurmountable political difficulties if national airspace was included.

GLOBAL CLIMATE CHANGE

At the Toronto Conference in June 1988, when the suggestion for adopting an international convention on the atmosphere by 1992 was made, many states, including the US, believed it to be an extremely early and, therefore, implausible target to achieve. Obviously, the political climate has changed a great deal since that time. In such fast-changing circumstances, it is easy to lose sight of one's perspective and to overlook some of the institutional innovations that took place during this time. Deserving special mention in this regard is the Intergovernmental Panel on Climate Change (IPCC).

The IPCC was established by the World Meteorological Organization (WMO) and the United Nations Environment Programme (UNEP) in November 1988. The Secretary-General of WMO, speaking at the opening session, emphasised the scientific nature of WMO's work in assisting 160 member countries in measuring, standardising, collecting and disseminating atmospheric data. Despite this scientific emphasis, and given the predicted global warming as one of the most important long-term challenges facing humanity, the Secretary-General said that the WMO cannot be a bystander while the consequences of the scientific studies are considered.[55] The Executive-Director of UNEP concurred with this assessment and hailed the formation and launching of the IPCC by UNEP and WMO as a milestone in global co-operation in addressing the challenge of climate change.

At its first session, held in Geneva in November 1988, the IPCC divided its work into three working groups.[56] They deal with scientific analysis,[57] socio- economic effects[58] and policy responses.[59] To improve co-ordination among the Working Groups, a Bureau was established, comprising of: the Chairman, Vice-Chairman and Rapporteur of IPCC; Chairmen of the three Working Groups; and Vice-Chairmen of the three Working Groups (two each from Working Groups I and II, and five from Working Group III). It was soon realised that not enough attention was being paid to ensure developing country participation. Therefore, a Special Committee on the Participation of Developing Countries was created, chaired by France.

Structured as it is, the IPCC has several features to distinguish itself from similar initiatives launched by international organisations in the past. Though the Terms of Reference emphasise the peer review process and though the participation in the various Working Groups and sub-groups is expected to be at a technical level, the participants were essentially either chosen by governments or are working with various governmental departments in the member countries. The *First Assessment Report*,[60] adopted at the Fourth Plenary of the IPCC, provided the conclusions of the three IPCC Working Groups, as well as the recommendations of the Special Committee on Developing Countries.

Prior to the adoption of this Report, Resolutions had been passed by WMO[61] and UNEP[62] to convene the first open-ended negotiating session on the framework convention on climate change. The first set of these meetings[63] took place in Geneva in September 1990 and was attended by over 70 countries.[64] The UNEP Executive-Director emphasised at this meeting that both he and the Secretary General of the WMO had been requested by their governing bodies 'to prepare for negotiations *now*'.[65] The UNEP/WMO note set out a number of points to be taken into consideration in the negotiation process, including: gases that ought to be included in atmospheric concentrations stabilisation and emission reductions targets, proposed dates, base year and criteria for calculating emission levels (per capita, per unit of GNP or GDP, according to the area of the country, its climatic conditions, the size of natural carbon sinks, energy consumption per production unit or a mixture of all these or other criteria).

Apart from the detail on what the convention should accomplish, there does seem to be agreement to a large measure that 'a meaningful legal act for adoption in 1992 should be a target'.[66] The question, however, is whether this should be accomplished at the expense of agreeing to an instrument of only declaratory character.

The resolve expressed by the executive heads of UNEP and WMO would be formally adopted by the Second World Climate Conference, which took place in Geneva in November 1990. The United States offered to host the first formal negotiating meeting. This brings the curtain down on at least the first part of an international process that is unprecedented for an environmental issue.

Several explanations were advanced on why the developing countries were active in the global climate change issue. The Special Committee of the IPCC on Developing Countries attempted to channel this interest in ways that would be conducive to long-term active involvement of developing countries. The task, as has been for some time now, is to translate what has now been accepted as concern on critical issues into specific policy measures. As Mostafa Tolba correctly pointed out, the three issues that would dwarf the entire negotiating process are: financial requirements, technology transfer and economic reforms.

CONCLUSION

The foregoing review clearly indicates that the approaches of the world community in addressing global environmental change vary from subject to subject. The approach followed in the global climate change issue is vastly different than that pursued in the case of the United Nations Conference on the Law of the Sea Convention. Yet, shades of the process adopted in the ozone negotiations are obvious enough. The success of the current climate change approach depends upon how and when governments agree on binding international legal agreements specifying clear commitments; defined responsibilities, targets and schedules; and workable equitable institutional mechanisms.[67] Once these issues are adequately addressed, the concerns raised with regard to the principle of Common Heritage of Mankind in this chapter will have been dealt with. Given the resolve of the world community to come up with legal instruments on global climate change for adoption at the United Nations Conference on Environment and Development to be held in Rio de Janeiro in 1992, the time leading up to that conference promises to be a period of intense international negotiations on this issue.

Notes

The author wishes to thank his research assistants, Dan Botsch of Boston University and Julie Williams at the Woods Hole Research Center, for their assistance in the preparation of this paper.

1. *India News* (Washington, DC: the Indian Embassy, Vol. 29, No. 4, 15–30 September 1990).
2. See Kilaparti Ramakrishna, 'The Emergence of Environmental Law in the Developing Countries: A Case Study of India', *Ecology Law Quarterly* (Vol. 12, No. 4, 1985) pp. 907–35.
3. See, for example, the Statement of Maneka Gandhi, Indian Minister of

State (Environment and Forests), at the White House Conference in Washington, DC, 17–18 April 1990, *India News* (Washington, DC: the Indian Embassy, April/May 1990). She emphasised the need to make available appropriate technologies and additional financial support and pointed out:

> When we cannot afford to set apart resources even to replace the unacceptable pesticides which affect the health of the people today, can we afford to make investments to avert a global problem which will be manifested 50 years hence, particularly when we are not contributing to the creation of the global problem?

4. Gro Harlem Brundtland, 'Our Common Future – A Climate for Change', *The Changing Atmosphere – Implications for Global Security: Conference Proceedings* (WMO/OMM, No. 710) pp. 16–23.
5. *Op. cit.*, in note 1.
6. *Ibid.*
7. See *op. cit.*, in note 4.
8. At the Ministerial Conference, 'Saving the Ozone Layer', hosted by the UK Prime Minister in London, March 1989.
9. See 'Remarks by Kilaparti Ramakrishna', in the session dealing with International Responsibility for Manmade Disasters, *Proceedings of the 81st Annual Meeting of the American Society of International Law*, pp. 336–8.
10. See *The State of India's Environment* (New Delhi, India: Centre for Science and Environment, 1985).
11. See *Official Text of the United Nations Convention on the Law of the Sea with Annexes and Index* (United Nations Publication Sales No. E.83.V5, 1983) Articles 212 and 222.
12. See, for example, the provisions of Part XI in relation to pollution from activities in the area.
13. UN Document, A/Conf. 62/L.61.
14. *Ibid.*
15. See, for example, *Ocean Development and International Law* (Vols 7–18); and *San Diego Law Review* (Vols. 8–23).
16. Louis B. Sohn, *Implications Of The Law Of The Sea Convention Regarding The Protection And Preservation Of The Marine Environment* (San Francisco, CA: Paper presented to the 18th Annual Conference on the Law of the Sea, Law of the Sea Institute and the University of San Francisco, 24 September 1984) p. 5.
17. *Op. cit.*, in note 11.
18. *Ibid.*, p. 74.
19. See, for example, Tanzania's comments on the document sponsored by the United Kingdom, UN Document, A/Conf. 12/C.3/L.24.
20. *Op. cit.*, in note 11.
21. See Fred Rucker, 'The Politics Of Ocean Pollution: The Third Law

of the Sea Conference and International Structures for Environmental Protection', *Boston College International and Comparative Law Journal* (Vol. 1, 1977) pp. 316–19.

22. *Ibid.*, pp. 313–14.
23. Sohn, *op. cit.*, in note 16.
24. *Ibid.*, p. 6.
25. *Ibid.* and *op. cit.*, in note 11, Articles 207(1), 208(3), 210(6) and 212(1).
26. For the text of the Vienna Convention for the Protection of the Ozone Layer, see *International Legal Materials* (Vol. 26) p. 1516.
27. For the text of the Montreal Protocol on Substances that Deplete the Ozone Layer, see *ibid.*, p. 1550.
28. See Richard Elliot Benedick, 'Building on the Vienna Convention: Lessons from "the Ozone Hole"', *EPA Journal* (March/April 1990) pp. 41–4.
29. While this question has not been directly addressed elsewhere, an interesting analysis, nevertheless, appears in Frederic L. Kirgis, Jr., 'Editorial Comment: Standing to Challenge Human Endeavors that Could Change the Climate', *American Journal of International Law* (Vol. 84, No. 2, April 1990) pp. 525–30.
30. Maurice Hope-Thompson, 'The Third World And The Law Of The Sea: The Attitude Of The Group Of 77 Toward The Continental Shelf', *Boston College Third World Law Journal* (Vol. 1, No. 1, 1980) p. 46.
31. *Ibid.*
32. D. H. N. Johnson, 'Air And Outer Space Law And The New International Economic Order', in *Air And Outer Space Law* (Thessaloniki, Greece: Thesaurus Acroasium of The Institute of Public International Law and International Relations of Thessaloniki, 1981) p. 390.
33. Burns H. Weston *et al.*, *Basic Documents In International Law And World Order* (St. Paul, MN: West Publishing Co., 1980) pp. 341–2.
34. Bradley Larschan and Bonnie C. Brennan, 'The Common Heritage of Mankind Principle in International Law', *Columbia Journal of Transnational Law* (Vol. 21, 1983) p. 305.
35. Christopher C. Joyner, 'Legal Implications of the Concept of the Common Heritage of Mankind', *International and Comparative Law Quarterly* (Vol. 35, 1986) pp. 191–2.
36. Weston, *op. cit.*, in note 33, pp. 341–2.
37. Carl Q. Christol, 'The International Law of Space Environment Resources', in *Air And Outer Space Law* (Thessaloniki, Greece: Thesaurus Acroasium of the Institute of Public International Law And International Relations of Thessaloniki, 1981) pp. 581–2.
38. See Johnson, *op. cit.*, in note 32, p. 398.
39. Leslie I. Tennen, 'Outer Space: A Preserve For All Humankind', *Houston Journal Of International Law* (Vol. 2, 1979) p. 154.
40. Boleslaw Adam Boczek, 'Ideology and the Law of the Sea: The

Challenge of the New International Economic Order', *Boston College International and Comparative Law Review* (Vol. 7, No. 1, 1984) p. 2.

41. Johnson, *op. cit.*, in note 32, p. 380.
42. *Ibid.*, p. 381.
43. *Ibid.*
44. See Joyner, *op. cit.*, in note 35, p. 196.
45. See Boczek, *op. cit.*, in note 40, pp. 1–2.
46. Christopher C. Joyner and Ethel R. Theis, 'The United States and Antarctica: Rethinking the Interplay of Law and Interests', *Cornell International Law Journal* (Vol. 20, 1987) p. 99.
47. Boczek, *op. cit.*, in note 40, p. 20.
48. *Ibid.*, pp. 49–50.
49. *Ibid.*, p. 13.
50. Hasjim Djalal, 'The Developing Countries and the Law of the Sea Conference', *Columbia Journal of World Business* (Vol. 15, 1980) p. 22.
51. Boczek, *op. cit.*, in note 40, p. 25.
52. *Ibid.*
53. Joyner, *op. cit.*, in note 35, pp. 198–9.
54. See Larschan and Brennan, *op. cit.*, in note 34: and Joyner, *op. cit.*, in note 35.
55. See *Report of the First Session of the WMO/UNEP Intergovernmental Panel on Climate Change (IPCC)* (Geneva, 9–11 November 1988).
56. According to the Terms of Reference, the reports of Working Groups should be written in such a way as to address the needs of policy-makers and non-specialists.
57. The Terms of Reference for Working Group I (chaired by the UK) are that it should consider

> factors affecting climate change, including greenhouse gases, responses to these factors of the atmosphere-ocean-land-ice system, assessment of current capabilities of modeling global and regional climate change and their predictability, past climate record and presently observed climate anomalies, projections of future climate and sea level and the timing of changes. The reports should identify the range of projections and their regional variations, gaps and uncertainties. The Working Group should consider including strategies to fill the knowledge gaps and reduce the uncertainties. A peer review should be incorporated in the preparation of the reports.

58. The Terms of Reference for Working Group II (chaired by the USSR) are that it should view

> the environmental and socio-economic impacts of climate change in an integrated manner. It should emphasise, *inter alia*, evaluation

> of the impacts on a regional/national scale of climate warming and sea level rise, the latter especially in the coastal and island areas, on agriculture, forestry, health, water resources and floods, droughts and desertification, energy and other sectors. The Working Group should include consideration of the impacts of a range of continuously changing climates. The reports should be peer-reviewed.

59. The Terms of Reference for Working Group III (chaired by the US) are that it should consider

> *inter alia*, forecasting and assessment of future emissions of greenhouse gases, impacts of changing technology, sources and sinks, adaptation to climate change, strategies to control or reduce emissions and other human activities that may have an impact on climate (*e.g.*, deforestation, changing land-use) and their social and economic implications, and legal matters.

60. See WMO/UNEP, *IPCC First Assessment Report: Overview* (31 August 1990).
61. See WMO Resolution 8 (EC-XLII), *Framework Convention on Climate Change.*
62. See UNEP Resolution SS.II/3, *Resolutions on the Intergovernmental Panel on Climate Change*; and *Framework Convention on Climate Change*, adopted at its 5th meeting on 3 August 1990.
63. See *Ad Hoc Working Group of Government Representatives to Prepare for Negotiations on a Framework Convention on Climate Change* (UNEP/WMO Prep./FCCC/L.1/Report).
64. It is revealing to see that among the attendees, there was an overwhelming number from the developing countries. This may augur well for ensuring both that developing country point of view is adequately taken into consideration and that whatever measures adopted to stabilise the composition of the atmosphere will truly have global support.
65. See *op. cit.*, in note 63. Emphasis in original.
66. *Ibid.*
67. *Ibid*, p. 2.

9 The Mirage of Eco-War: The Weak Relationship among Global Environmental Change, National Security and Interstate Violence

Daniel Deudney

INTRODUCTION

Like all biological organisms, humans are vitally dependent upon their physical environment. Since the emergence of human life on earth, humans have been able simply to take as given the presence of some environmental conditions – for example, clean air and shielding from ultraviolet radiation – that are now in jeopardy. Other environmental elements – particularly fertile soil, water and earth minerals – have been subject to intense, often violent, intergroup competition.[1] Over the last two centuries, the explosive progress in science and technology and the emergence of societies of unprecedented wealth seemed to have loosened the iron grip of natural scarcity upon human life. In the last several decades, however, alarming evidence has accumulated that human activities have begun to cause significant changes in the earth's life support system.

Because of their planetary scope and potentially catastrophic consequences, recent discoveries about alterations in the chemical composition of the earth's atmosphere caused by human activity are particularly alarming.[2] There is a scientific consensus that human activities are increasing the concentration of several gases (most importantly, carbon dioxide, chlorofluorocarbons [CFCs] and methane) in the atmosphere. It is generally accepted that CFC release has led to a depletion of stratospheric ozone, which could have significant effects upon human populations, animals, material objects and the natural environment. Further, there is an emerging scientific consensus that these changes in the atmosphere's chemistry are also likely to produce an enhanced 'greenhouse effect' – a rise in atmospheric temperatures. Although the exact timing and scope of global atmospheric warming is subject to scientific dispute, a substantial number of scientists believe that current and projected accumulations in such 'greenhouse gases' will begin to cause a wide range of major and far-reaching effects, including a rise in sea-level with consequent flooding of low-lying coastal regions, alterations in rainfall patterns with consequent droughts, floods and diminished agricultural output, and sudden changes in habitats causing massive die-offs of plant and animal life and extensive species extinction. Given these alarming projections and trends, environmental issues are likely to become an increasingly important dimension of political life at all levels – locally, nationally, as well as internationally. The sudden emergence of global environmental change upon the world political agenda raises a host of important practical and theoretical questions for political scientists. In a recent issue of *Scientific American*, William Ruckelshaus, the former Administrator of the US Environmental Protection Agency, asserts that the achievement of sustainability 'would be a modification of society comparable in scale to only two other changes: the agricultural revolution of the Late Neolithic period and the Industrial Revolution of the past two centuries'.[3] Considering the difficulty humans and their institutions have in adjusting rapidly and peacefully to major changes, Ruckelshaus' prediction bodes ill for human survival. Hopefully, his comments exaggerate the changes required. However, if change of this magnitude is needed for human survival, then political theorists are invited to envision worlds that are as different from contemporary society as contemporary society is from either earlier agricultural societies or still earlier societies of hunters and gatherers.

An impending change of this magnitude amounts to an invitation to call the entire conceptual landscape into doubt, including the most basic assumptions of modern life. If in fact a shift of this magnitude (and one compressed into such a short time-frame) is imminent, then vivid political imaginations are required to conceive of different types of associations and institutions. For conceptualising such radically different worlds, science fiction writers and futurists may be better equipped than social scientists.[4] Adequately answering questions of such scope and profundity will require the work of many minds, perhaps stretching over several generations. This task will involve not just political scientists, but will require contributions from many disciplines in the natural sciences, social sciences and humanities.

Since the *state*[5] and the *nation-state* occupy such central positions in the contemporary world political scene, it is logical to start by analysing the connections between this pervasive institution and global environmental change. As Stanley Hoffmann once suggested, the writings of international relations scholars tend to resemble entries into the contest for the best essay on the 'persistence or demise of the realist paradigm' and the nation-state system.[6] Unfortunately, resolving the status of the state and the system of states in contemporary world politics is no simple matter. The essence and extent, even the existence, of the state is a highly contested issue. As James Rosenau recently observed: Whether the state is thought to be widening its competence, a withering colossus or merely weathering change depends upon which of several wavering concepts one employs.[7]

Although the elusive modern state is involved in virtually all aspects of political life, its oldest and in many ways still most defining activities concern war, violence and protection. Contemporary states, like their predecessors, go to great effort to deploy military forces deemed sufficient to provide (at least) the security of their territory. Although the incidence of violence, particularly among industrialised states, has fallen since the end of World War II, states continue to pay close attention to their military security. The state system and the war system are not identical, but the war system might be thought of as the most fundamental and original aspect of the state-system. The question posed before us therefore is: How will global environmental change affect national security and the frequency of interstate violence? Here I advance three claims. First, environmental degradation stemming from global environmental change has little in common with national-security-from-violence. Second, declaring such environmental changes a threat to national security (as many environmental activists do) in order to mobilise action may be counterproductive and produce undesirable side-effects. And third, global environmental change is unlikely to produce interstate war.

THE WEAK ANALYTICAL LINKS BETWEEN GLOBAL ENVIRONMENTAL CHANGE AND NATIONAL SECURITY

One striking feature of the growing discussion of environmental issues in the United States is the attempt by many liberals, progressives and environmentalists to employ language traditionally associated with violence and war to understand environmental problems and to motivate action. Lester Brown, Jessica Tuchman Mathews, Michael Renner and others have proposed 'redefining national security' to encompass resource and environmental threats.[8] More broadly, Richard Ullman and others have proposed 'redefining security' to encompass a wide array of threats, ranging from earthquakes to environmental degradation.[9] Hal Harvey has proposed the concept of 'natural security',[10] and US Senator Albert Gore has spoken extensively in favour of thinking of the environment as a 'national security issue'.[11] During the renewed Cold War tensions of the late 1970s and early 1980s, such concepts were advanced to prevent an excessive focus on military threats. As the Cold War winds down, such links are increasingly popular among national security experts and organisations looking for new missions, as exemplified by US Senator Sam Nunn's initiative to spend US$200 million for efforts in environmental monitoring and research.[12] Due to the interest and support of several foundations, numerous conferences and researchers are addressing issues of 'environmental security'. (In characterising ecological decay as a 'national security' problem, these analysts naturally do not give exclusive attention to the problems posed by global environmental change, so this debate encompasses global environmental change but is not unique to it.)

Historically, conceptual ferment of this sort has often accompanied important changes in politics.[13] New phrases are coined and old terms are appropriated for new purposes. Epochal developments like the emergence of capitalism, the growth of democracy and the end of slavery were accompanied by shifting, borrowing and expanding political language. The wide-ranging contemporary conceptual ferment in the language used to understand and act upon environmental problems is therefore both a natural and an encouraging development.

But not all neologisms and linkages are equally plausible or useful. Until this recent flurry of reconceptualising, the concept of 'national security' (as opposed to national interest or well-being) had been centred upon *organised violence*.[14] As is obvious to common sense and as Hobbes argued with such force, security from violence is a primal human need, because loss of life prevents the enjoyment of all other goods. Of course, various resource factors, such as access to fuels and ores, were understood as to contribute to states' capacities to wage war and achieve security from violence.

Before either 'expanding' the concept of 'national security' to encompass both environmental and violence threats, or 're-defining' 'national security' or 'security' to refer mainly to environmental threats, it is worth examining just how much the national pursuit of security from violence has in common with environmental problems and their solutions.

Military violence and environmental degradation are linked directly in at least three major ways. First, the pursuit of national-security-from-violence through military means consumes resources (fiscal, organisational and leadership) that could be spent on environmental restoration. Since approximately one trillion US dollars is spent worldwide on military activities, substantial resources are involved. However, this relationship is not unique to environmental concerns, and unfortunately there is no guarantee that money saved from military expenditures would be spent on environmental restoration. Nor is it clear that the world cannot afford environmental restoration without cutting military expenditures.

Second, war is directly destructive of the environment.[15] In ancient times, the military destruction of olive groves in Mediterranean lands contributed to the long-lasting destruction of the lands' carrying capacities. More recently, the United States' bombardment and use of defoliants in Indochina caused significant environmental damage. Further, extensive use of nuclear weapons could have significant impacts on the global environment, including altered weather (that is, 'nuclear winter') and further depletion of the ozone layer. Awareness of these environmental effects has played an important role in mobilising popular resistance to the arms race and in generally de-legitimising the use of nuclear explosives as weapons.

Third, preparation for war causes pollution and consumes significant quantities of resources. In both the United States and the Soviet Union, significant quantities of radioactive waste have been produced as a by-product

TABLE 9.1 *Conceptual and Organisational Mismatches*

	Conventional National Security	*Global Habitability*
type of threat	– violent death – destruction of propety – loss of independence	– wide range of harms: aesthetics, disease, natural integrity
source of threat	– mainly outside – other states armed with weapons	– both inside and outside – wide range of sources: individuals, corporations, governments
degree of intentionality	– direct and high – 'accidental' war	– largely unintentional side-effects of routine activities – accidental spill, release and so on
types of organisations involved	– specialised – secretive – removed from civil society	– all sizes – *in situ* change of many mundane activities: land use, waste treatment, farming, factory design

of the nuclear arms race, and several significant releases of radiation have occurred – perhaps most disastrously when a waste dump at a Soviet nuclear weapons facility exploded and burned, spreading radioactive materials over a large area near the Urals.[16] Military activities have also produced significant quantities of toxic wastes.

In short, war and the preparation for war are clearly environmental threats and consume resources that could be used to ameliorate environmental degradation. In effect, these environmental impacts mean that the war system has costs beyond the intentional loss of life and destruction. Nevertheless, most of the world's environmental degradation is not caused by war and preparation for war. Completely eliminating the direct environmental effects of the war system would leave most environmental degradation unaffected. Most of the causes and most of the cures of environmental degradation must be found outside the domain of the traditional national security system related to violence.

The war system is a definite but limited environmental threat, but in what ways is the environmental degradation stemming from global environmental change a threat to 'national security'? Making such an identification can be useful if the two phenomenon – security from violence and security from environmental threats – are similar. Upon closer examination, they have little in common, making such linkages largely useless for analytical and conceptual purposes. Four major dissimilarities, listed in Table 9.1, deserve mention.

First, environmental degradation and violence are very different types of threats. Both violence and environmental degradation may kill people and may reduce human well-being, but not all threats to life and property are threats to security. Disease, old age, crime and accidents routinely destroy life and property, but we do not think of them as 'national security' threats or even threats to 'security'. (Crime is a partial exception, but crime is a 'security' threat at the individual level, because crime involves violence.) And when an earthquake or hurricane strikes with great force, we speak about 'natural disasters' or designate 'national disaster areas', but we do not speak about such events threatening 'national security'. If everything that causes a decline in human well-being is labelled a 'security' threat, the term loses any analytical usefulness and becomes a loose synonym of 'bad'.

Second, the scope and source of threats to environmental well-being and national-security-from-violence are very different. There is nothing about the problem of environmental degradation which is particularly 'national' in character. The problems of global environmental change do not afflict just one nation-state, but rather, they ignore the borders of the nation-state altogether. It would, however, be equally misleading even to call this problem solely 'international'. Many perpetrators and victims are within the borders of one nation-state. Individuals, families, communities, other species and future generations are harmed. A complete collapse of the biosphere would surely destroy 'nations' as well as everything else, but there is nothing distinctively national about either the causes, the harms or the solutions that warrants us giving such privileged billing to the 'national' grouping.

A third misfit between environmental well-being and national-security-from-violence stems from the differing degrees of *intention* involved. Violence threats involve a high degree of intentional behaviour. Organisations are mobilised, weapons procured and wars waged with relatively definite aims in mind. Environmental degradation, on the other hand, is largely unintentional, the side-effects of many other activities. No one really sets out with the aim of harming the environment (with the so far limited exception of environmental modification for military purposes).

Fourth, organisations that provide protection from violence differ greatly from those in environmental protection. National-security-from-violence is conventionally pursued by organisations with three distinctive features. First, military organisations are secretive, extremely hierarchical and centralised, and normally deploy vastly expensive, highly specialised and advanced technologies. Second, citizens typically delegate the goal of achieving national security to remote and highly specialised organisations that are far removed from the experience of civil society. And third, the members of the specialised professional group staffing these national security organisations are trained in the arts of killing and destroying.

In contrast, responding to the environmental problem requires almost exactly opposite approaches and organisations. Certain aspects of virtually all mundane activities – for example, house construction, farming techniques, sewage treatment, factory design and land use planning – must be reformed. The routine everyday behaviour of practically everyone must be altered. This requires behaviour modification *in situ*. The professional ethos of environmental restoration is husbandmanship – more respectful cultivation and protection of plants, animals and the land.

In short, national-security-from-violence and environmental habitability have little in common. Given these differences, the rising fashion of linking them risks creating a conceptual muddle rather than a paradigm or world view shift – a *de-definition* rather than a *re-definition* of security. If we begin to speak about all the forces and events that threaten life, property and well-being (on a large-scale) as threats to our national security, we shall soon drain the term of any meaning. All large-scale evils will become

TABLE 9.2 *Associated Mindsets*

	Conventional National Security		*Global Habitability*
1.	urgency/crisis; make sacrifices; no expense is too great	1.	urgency/crisis?; make sacrifices; no expense is too great or frugality?
2.	worse case scenarios as basis for planning	2.	worse case scenarios as basis for planning
3.	mainly zero-sum	3.	common benefits
4.	short-term time horizons	4.	long-term time horizons
5.	nationalism; us *vs*. them	5.	'the enemy is us'; polluters in other countries are threat; pollution in other countries is threat

threats to national security. To speak meaningfully about actual problems, we shall have to invent new words to do the job previously performed by the old spoiled ones.

THE RISKS IN HARNESSING THE RHETORICAL AND EMOTIONAL APPEALS OF NATIONAL SECURITY FOR ENVIRONMENTAL RESTORATION

Confronted with these arguments, the advocate of treating environmental degradation as a national security problem might retort:

> Yes, some semantic innovation without much analytical basis is occurring, but it has a sound goal – to get people to react as urgently and effectively to the environmental problem as they have to the national-security-from-violence problem. If people took the environmental problem as seriously as, say, an attack by a foreign power, think of all that could be done to solve the problems!

In other words, the aim of these new links is not primarily descriptive, but polemical. It is not a claim about fact, but a rhetorical device designed to stimulate action. Like William James, these environmentalists hope to find a 'moral equivalent to war' to channel the energies behind war into constructive directions.

But before harnessing the old horse of national security to pull the heavy new environmental wagon, prudence demands a closer look at its temperament. Potential disadvantages as well as the apparent advantages should be considered. The sentiments associated with national security are so powerful, because they relate to war. As the historian Michael Howard has observed:

> Self-consciousness as a Nation implies, by definition, a sense of differentiation from other communities, and the most memorable incidents in the group memory usually are of conflict with, and triumph over, other communities. It is in fact very difficult to create national self-consciousness *without* a war.[17]

Such sentiments often run deep, but they also run in directions different from those needed for successful environmental politics. If the emotional appeals of national security can somehow be connected to environmental issues, then it is also possible that other, less benign, associations may be transferred. To make a balanced assessment of such 'mindset' factors is (like most of the things that matter in politics) to grapple with intangibles not easily quantified – but vital nevertheless (see Table 9.2).

At first glance, the most attractive feature of linking fears about environmental threats with national security mentalities is the sense of urgency engendered, and the corresponding willingness to accept great personal sacrifice. If in fact the basic habitability of the planet is being undermined, then it stands to reason that some crisis mentality is needed. Unfortunately, it may be difficult to engender a sense of urgency and a willingness to sacrifice for extended periods of time. Crises call for resolution, and the patience of a mobilised populace is rarely long. For most people, exertion in a crisis is motivated by a desire to return to normalcy, for the problem to be resolved once and for all. Such a cycle of arousal and somnolence is not likely to make much of a contribution to establishing sustainable patterns of environmentally sound behaviour. The limits of attempting to create a 'moral equivalent of war' mentality to reform long-term resource-use patterns was vividly demonstrated during the late 1970s when US President Carter's effort to mobilise public awareness and forge a public consensus about energy conservation produced a flurry of activity that subsided once the immediate symptoms of the problem receded. Furthermore, 'crash' solutions are often bad ones – more expensive, more oppressive and more poorly designed than typical government programmes, as exemplified by such US white elephants as the proposed synfuels programme, the 'energy mobilization board' and the Byzantine system of price controls spawned by the 'energy security' crisis.

A second apparently valuable similarity between the national security mentality and the environmental problem is the tendency to use worse case scenarios as the basis for planning. However, the extreme conservatism of military organisations in responding to potential threats is not unique to them. The insurance industry is built around preparations for the worst possibilities, and many fields of engineering, such as aeronautical design and nuclear power plant regulation, routinely employ extremely conservative planning assumptions. These can serve as useful models for improved environmental policies.

Third, the conventional national security mentality and its organisations are deeply committed to zero-sum thinking. 'Our' gain is 'their' loss. Trust between national security organisations is extremely low. The prevailing assumption is that everyone is a potential enemy, and that agreements mean little unless congruent with immediate interests. If the Pentagon had been put in charge of negotiating an ozone layer protocol, we might still be stockpiling chlorofluorocarbons as a bargaining chip.

Fourth, conventional national security organisations have short time horizons. The pervasive tendency for national security organisations to discount the future and pursue very near-term objectives is a poor model for environmental problem solving.

Finally, and perhaps most importantly, is the fact that the 'nation' is not an empty vessel or blank-slate waiting to be filled or scripted, but is instead profoundly linked to war and 'us *vs.* them' thinking. The tendency for people to identify themselves with various tribal and kin groupings is as old as humanity. In the last century and a half, however, this sentiment of nationalism, amplified and manipulated by mass media propaganda techniques, has been an integral part of totalitarianism and militarism. Nationalism means a sense of 'us *vs.* them', of the insider *vs.* the outsider, of the compatriot *vs.* the alien. The stronger the nationalism, the stronger this cleavage, and the weaker the transnational bonds. Nationalism reinforces militarism, fosters prejudice and discrimination, and feeds the quest for 'sovereign' autonomy. These fundamental features of nationalism are often forgotten by intellectuals who often reject such sentiments as irrational. National security thinking and action is all premised upon a relatively sharp distinction between 'us' and 'them', between friend and foe. Of course, taking the war and the 'us *vs.* them' out of nationalism is a noble goal. But this may be like taking sex out of 'rock and roll', a project whose feasibility declines when one remembers that 'rock and roll' was originally coined as a euphemism for sex.

In contrast, in the environmental sphere 'we' – not 'they' – are the 'enemy', as Pogo reminds us. As noted earlier, existing 'us *vs.* them' groupings in world politics match very poorly the causal lines of environmental degradation. At its most basic level, the environmental problem asks us to redefine who 'us' encompasses. Coping with global problems and new forms of interdependence require replacing or supplementing national identity with other forms of group identity. Intense nationalism directly conflicts with the globalism that has been one of the most important insights of environmentalism. If in fact resolution of the global environmental problem, and particularly the global warming problem, requires great, even unprecedented, types of international co-operation, then nationalist sentiment and identification is a barrier to be overcome. Thus, thinking of national security as an environmental problem risks undercutting both the globalist and common fate understanding of the situation and the sense of world community that may be necessary to solve the problem. In short, it seems doubtful that the environment can be wrapped in national flags without undercutting the 'whole earth' sensibility at the core of environmental awareness.

If pollution comes to be seen widely as a national security problem, there is also a danger that the citizens of one country will feel much more threatened by the pollution from other countries than by the pollution created by their fellow citizens. This could increase international tensions and make international accords more difficult to achieve, while diverting attention from internal clean up. Citizens of the United States, for example, could become much more concerned about deforestation in Brazil than in reversing the centuries of North American deforestation. Taken to an absurd extreme – as national security threats sometimes are – seeing environmental degradation in a neighbouring country as a national security threat could trigger various types of interventions – a new imperialism of the strong against the weak.

Instead of linking 'national security' to the environmental problem, environmentalists should emphasise that the environmental crisis calls into question the national grouping and its privileged status in world politics. The environmental crisis is not a threat to national security, but it does challenge the utility of thinking in 'national' terms.

Proponents of thinking of environmental degradation as a national security problem may react to my objections with the following argument:

> Yes, nationalist, statist and militarist features of 'national security' thinking are evil and rampant, but we globalists now know that security

> in the old and limited sense of protection from organised violence can now only be achieved in common. Henceforth, both the environmental and the violence threats to our security can only be achieved in 'common'.

Such a response reveals a key feature of the political sociology of environmentalism, namely, that most people who take environmentalists seriously adhere to a 'common security' understanding of the organised violence problem. The notion of 'common security' has enjoyed great vogue among progressive intellectuals since the beginning of this century, particularly in English-speaking and Nordic countries. With the coming of nuclear explosives, the 'common security' characterisation of the violence problem would seem to take on the air of the obvious. But the concept of 'common security' is not widely accepted in principle outside certain progressive circles, despite the sound reasoning behind it. Nor is it significantly reflected in the actual practice of state violence-security organisations. Indeed, the conceptual apparatus of 'common security' thinking has in large measure been fashioned as a fundamental alternative to mainstream 'national security' thinking and practice. The phrase 'common security' as applied to the organised violence problem is still 'revolutionary' and 'radical' in the sense that it would alter beyond recognition the practice and mindset of protection-from-violence organisations. All the talk about a 'nuclear revolution' should not obscure the fact that we are in a 'revolutionary predicament' rather than in a 'post-revolutionary state'. Given this reality, it is premature to characterise environmental degradation as a threat to 'security' until we can be more confident that people do in fact think 'common' whenever they hear 'security'.

Fortunately, environmental awareness need not depend upon co-opted national security thinking. Integrally woven into ecological awareness are a powerful set of values and symbols – ranging from human health and property values to beauty and concern for future generations. These values and symbols draw upon basic human desires and aspirations and are powerful motivators of human action. Far from needing to be bolstered by national security mindsets – which in the wake of the nuclear revolution are themselves of dubious value, even on their own terms – this 'green' sensibility can make strong claim to being the master metaphor for an emerging post-industrial civilisation. Instead of attempting to gain leverage by appropriating national security thinking, environmentalists should continue to develop and disseminate this rich emergent world view.

ENVIRONMENTAL DEGRADATION AND INTERSTATE WAR

Many people are drawn to calling environmental degradation a national security problem, in part because they expect this phenomenon to stimulate interstate conflict and even violence. States often fight over what they value, particularly if related to 'security'. If states begin to be much more concerned with resources and environmental degradation, particularly if they think environmental decay is a threat to their 'national security', then states may well fight resource and pollution wars. Much of the recent literature on the impacts of global environmental change upon world politics posits conflictual and violent outcomes.[18] As Arthur Westing has noted that

> global deficiencies and degradation of natural resources, both renewable and non-renewable, coupled with the uneven distribution of these raw materials, can lead to unlikely – and thus unstable – alliances, to national rivalries, and, of course, to war.[19]

In emphasising such outcomes, environmental analysts essentially assume – along with realist international relations scholars – that international political life follows a conflictual and violent pattern.

Social science is far from achieving a satisfactory understanding of human conflict and war, and environmental problems are imperfectly understood as well. Therefore, analysing the prospects for violent outcomes is an uncertain undertaking at best.[20] There are more than 150 independent states, and resource and environmental problems are diverse, therefore any generalisation will surely have important exceptions. However, a rough assessment of the prospects for resource and pollution wars can be made by examining five of the most frequently mentioned ways in which environmental deterioration could possibly cause interstate conflicts and war. In analysing such possible linkages, it is important to keep in mind that conflict and change are endemic in human affairs, but that they only lead to interstate violence under exceptional circumstances.

Few ideas seem more intuitively sound than the notion that states will begin fighting each other as the world runs out of useable natural resources. The popular metaphor of a lifeboat adrift at sea with declining supplies of clean water and rations suggests there will be fewer and fewer opportunities for positive-sum gains between actors. Many ideas about resource war are derived from the cataclysmic world wars of the first half of the twentieth century. Influenced by geopolitical theories that emphasised the importance of land and resources for Great Power status, Hitler in significant measure fashioned Nazi war aims to achieve resource autonomy.[21] The aggression of Japan was directly related to resource goals: lacking indigenous fuel and minerals, and faced with a slowing tightening embargo by the Western colonial powers in Asia, the Japanese invaded Southeast Asia for oil, tin and rubber.[22] Although the United States had a richer resource endowment than the Axis powers, fears of shortages and industrial strangulation played a central role in the strategic thinking of American elites about world strategy.[23] And during the Cold War, the presence of natural resources in the Third World helped turn this vast area into an arena for East-West conflict.[24] Given this record, the scenario of conflicts over resources playing a powerful role in shaping international order should be taken seriously.

There are, however, three strong reasons for concluding that the familiar scenarios of resource war are of diminishing plausibility for the foreseeable future. First, the robust character of the world trade system means that states no longer experience resource dependency as a major threat to their military security and political autonomy.[25] During the 1930s, the world trading system collapsed, driving states to pursue autarkic economies. In contrast, the resource needs of contemporary states are routinely met without territorial control of the resource source, as Ronnie Lipschutz has recently shown.[26]

Second, the prospects for resource wars are diminished by the growing difficulty for states to exploit foreign resources through territorial conquest. Although the invention of nuclear explosives has made it easy and cheap to annihilate humans and infrastructure in extensive areas, the spread of small arms and national consciousness has made it very costly for an invader, even one equipped with advanced technology, to subdue a resisting population – as France discovered in Indochina and Algeria, the United States in Vietnam and the Soviet Union in Afghanistan.[27] Schemes of resource imperialism are now more appealing to romantic militarists than practical policy-makers. At the lower levels of violence capability – those that matter for conquering and subduing territory – the Great Powers have lost an effective military superiority and are unlikely soon to regain it.

Third, the world is entering what H. E. Goeller and Alvin M. Weinberg have called the 'age of substitutability', in which industrial civilisation is increasingly capable of taking earth materials such as iron, aluminum, silicon and hydrocarbons (which are ubiquitous and plentiful) and fashioning them into virtually everything needed.[28] The most striking manifestation of this trend is that prices for virtually every raw material have been stagnant or falling for the last several decades, despite the continued growth in world output. In contrast to the expectations voiced by many during the 1970s – that resource scarcity would drive up commodity prices to the benefit of Third World raw material suppliers – prices have fallen, with disastrous consequences for Third World development.

The scenarios of resource conflict most frequently mentioned by environmental analysts posit that disputes over water supplies will become acute as rainfall and runoff patterns are altered by global warming. A large number of rivers cross the territory of one or more states, and water is already in increasingly short supply in many arid regions of the world. As a general rule, however, it seems less likely that conflicts over water will lead to interstate war than that the development of jointly owned water resources will reinforce peace. Exploitation of water resources typically requires complex and expensive civil engineering systems, such as dams and pipelines, and such infrastructure is extremely vulnerable to disruption and destruction, as Amory and Hunter Lovins have documented.[29] Large dams, like nuclear power plants, are 'potential weapons in the hands of the enemy'. Thus the development of water resources creates a mutual hostage situation, reducing greatly the incentives for states to employ violence to resolve conflicts. Furthermore, there is evidence that the development of water resources by antagonistic neighbours creates a network of common interests.

In a second scenario, increased interstate violence results from internal turmoil caused by declining living standards. Many commentators on the environmental crisis emphasise that the basic source of environmental distress is the modern success in producing wealth. Some maintain that new technology, institutional reform, new attitudes and more efficient capital investment can largely solve the environmental problem without sacrificing high standards of living. Others, however, are more pessimistic, asserting that the great levels of wealth produced (if unevenly distributed) since the industrial revolution cannot be sustained. In this view, the only way to prevent ecological collapse (and the resulting economic collapse) is for people to accept radically lower standards of living.

Ramifications of economic stagnation upon politics and society could well be major and largely undesirable. Although the peoples of the world could perhaps live peacefully at lower standards of living, reductions of expectations to conform to these new realities will not come easily. Faced with declining living standards, groups at all levels of affluence can be expected to resist this trend by pushing the deprivation upon other groups. Class relations would be increasingly 'zero sum games', producing class war and revolutionary upheavals. Faced with these pressures, liberal democracy and free market systems would increasingly be replaced by authoritarian systems capable of maintaining minimum order.[30]

The consequences of these domestic changes on the international system may be increased conflict and war. If authoritarian regimes are more war-prone because of their lack of democratic control, and if revolutionary regimes are more war-prone because of their ideological fervour and lack of socialisation into international norms and processes, then a world political system containing more such states is likely to be an increasingly violent one. The historical record from previous economic depressions supports the general proposition that widespread economic stagnation and unmet economic expectations contribute to international conflict.

Although initially compelling, this scenario has flaws as well. First, the pessimistic interpretation of the relationship between environmental sustainability and economic growth is arguably based on unsound economic theory. Wealth formation is not so much a product of cheap natural resource availability as of capital formation via savings and more efficient ways of producing. The fact that so many resource-poor countries, like Japan, are very wealthy, while many countries with more extensive resource endowments are poor, suggests that there is no clear and direct relationship between abundant resource availability and national wealth. Environmental constraints require an end to economic growth based on increasing raw material through-puts, rather than an end to growth in the output of goods and services.

Second, even if economic decline does occur, interstate conflict may be dampened, not stoked. In the pessimistic scenario, domestic political life is an intervening variable connecting environmentally induced economic stagnation with interstate conflict. How societies respond to economic decline may in large measure depend upon the rate at which such decline occurs. An off-setting factor here is the possibility that as people get poorer, they will be less willing to spend increasingly scarce resources for military capabilities. In this regard, the experience of economic depressions over the last two centuries may not be relevant, because such depressions were characterised by under-utilised production capacity and falling resource prices. In the 1930s, increased military spending had a stimulative effect, but in a world in which economic growth had been retarded by environmental constraints, military spending would exacerbate the problem.

Third, environmental degradation may affect interstate relations in such a way as to cause war by altering the relative power capacities of states. Alterations in relative power position can contribute to wars, either by tempting the newly strengthened states to aggress upon the newly weakened ones or by leading the newly weakened ones to attack and lock in their power position relative to their neighbours before their power ebbs any further. Support for such scenarios can be drawn from history, and international relations scholars have extensively studied such phenomenon.[31]

Nevertheless, such alterations in the relative international power potential of states might not lead to war as readily as the lessons of history suggest, because economic power and military power are perhaps not as tightly coupled as in the past. The relative economic power positions of major states, such as Germany and Japan, have altered greatly since the end of World War II. But these changes, while requiring many complex adjustments in interstate relations, have not been accompanied by war or threat of war. In the contemporary world, whole industries rise, fall and re-locate, often causing quite substantial fluctuations in the economic well-being of regions and peoples, but wars do not result. What reason is there for believing that changes in relative wealth and power position caused by the uneven impact of environmental degradation would be different in their effects?

Part of the reason for this loosening of the economic-military link has been the nuclear revolution, which has made it relatively cheap for the leading states to sustain a mutual kill capacity. Given that the superpowers field massively oversufficient nuclear forces at the cost of a few per cent of their GNP, environmentally-induced economic decline would have to be extreme before their ability to field a minimum nuclear deterrent would be jeopardised.

Environmental degradation in a country or region could become so extreme that the basic social and economic fabric comes apart. Should some areas of the world suffer this fate, the impact of this outcome on international order may not, however, be very great. If a particular country, even a large one like Brazil, were tragically to disintegrate, among the first casualties would be the capacity of the industrial and governmental structure to wage and sustain interstate conventional war. As Bernard Brodie observed about the modern era, 'the predisposing factors to military aggression are full bellies, not empty ones'.[32] The poor and wretched of the earth may be able to deny an outside aggressor an easy conquest, but they are themselves a minimal threat to outside states. Offensive war today requires complex organisational skills, specialised industrial products and surplus wealth.

In today's world everything is connected, but not everything is tightly coupled. Regional disasters of great severity may occur, with scarcely a ripple in the rest of the world. After all, Idi Amin drew Uganda back into savage darkness, the Khmer Rouge murdered an estimated two million Cambodians and the Sahara has advanced across the Sahel without the economies and political systems of the rest of the world being much perturbed. Indeed, many of the world's citizens did not even notice.

A fourth possible route from environmental degradation to interstate conflict and violence involves pollution across interstate borders. It is easy to envision situations in which country A dumps an intolerable amount of pollution on a neighbouring country B (which is upstream and upwind), causing country B to attempt to pressure and coerce country A into eliminating its offending pollution. We can envision such conflict of interest leading to armed conflict.

Fortunately for interstate peace, strongly asymmetrical and significant environmental degradation between neighbouring countries is relatively rare. Probably more typical is the situation in which activities in country A harm parts of country A and country B, and in which activities in country B also harm parts of both countries. This creates complex sets of winners and losers, and thus establishes a complex array of potential intrastate and interstate coalitions. In general, the more such interactions are occurring, the less likely it is that a persistent, significant and highly asymmetrical pollution 'exchange' will result. The very multitude of interdependency in the contemporary world, particularly among the industrialised countries, makes it unlikely that intense cleavages of environmental harm will match interstate borders, and at the same time not be compensated and complicated by other military, economic or cultural interactions. Resolving such conflicts will be a complex and messy affair, but the conflicts are unlikely to lead to war.

Finally, there are conflict potentials related to the global commons. Many countries contribute to environmental degradation and many countries are harmed, but since the impacts are widely distributed, no one country has an incentive to act alone to solve the problem. Solutions require collective-action, and with collective action comes the possibility of the 'free rider'. In the case of a global agreement to reduce carbon dioxide emissions, if one significant polluter were to resist joining the agreement, with the expectation that the other states would act to reduce the harms to a tolerable level, the possibility would thus arise that those states making the sacrifices to clean up the problem would attempt to coerce the 'free rider' into making a more significant contribution to the effort.

It is difficult to judge this scenario, because we lack examples of this phenomenon on a large scale. 'Free-rider' problems may generate severe conflict, but it is doubtful that states would find military instruments useful for coercion and compliance. If, for example, China decided not to join a global climate agreement, it seems unlikely that the other major countries would really go to war with China over this. In general, any state sufficiently industrialised to be a major contributor to the carbon dioxide problem will also present a very poor target for military coercion.

To summarise, the case for thinking that environmental degradation will cause interstate violence is much weaker than commonly thought. In part this is because of features of the international system – particularly the hypertrophy of violence capability available to states – that have little to do directly with environmental matters. Although many analogies for such conflict draw from historical experience, they fail to take into account the ways in which the current interstate system differs from earlier ones. Military capability sufficient to make aggression prohibitively costly has become widely distributed, making even large shifts in the relative power potential of states less likely to cause war. Interstate violence seems to be poorly matched as a means to resolve many of the conflicts that might arise from environmental degradation. The vitality of the international trading system and the more general phenomenon of complex interdependency also militate against violent interstate outcomes. The result is a world system with considerable resiliency and 'rattle room' to weather significant environmental disruption without significant violent interstate conflict.

CONCLUSION

Global environmental change presents challenges of far-reaching significance for human societies everywhere. But this challenge has little to do with the national-security-from-violence problem that continues to plague human political life. Not only is there little in common between the causes and solutions to these two problems, but the nationalist and militarist mindsets closely associated with 'national security' thinking directly conflict with the core of the environmentalist world view. Harnessing these sentiments for a 'war on pollution' is a dangerous and probably self-defeating enterprise. And fortunately, the prospects for resource and pollution wars are not as great as often conjured by environmentalists.

The recourse to national security paradigms to conceptualise the environmental problem represents a profound and disturbing failure of political imagination and historical awareness. If the nation-state enjoys a more prominent status in world politics than its competence and accomplishments warrant, then it makes little sense to emphasise the links between it and the emerging problem of the global habitability.[33] Nationalist sentiment and the war system have a long-established logic and staying power that are likely to defy any rhetorically conjured 're-direction' towards benign ends. The movement to preserve the habitability of the planet for future generations must directly challenge the tribal power of nationalism and the chronic militarisation of public discourse. Environmental degradation is not a threat to national security. Rather, environmentalism is a threat to 'national security' mindsets and institutions. For environmentalists to dress their programmes in the blood-soaked garments of the war system betrays their core values and creates confusion about the real tasks at hand.

Notes

Earlier versions of this chapter were presented at: the American Academy of Arts and Sciences in Boston, MA in November 1989; the Center for Energy and Environmental Studies at Princeton University, Princeton, NJ in December 1989; the Five College Peace Consortium in January 1990; and the Center for International Cooperation and Security Studies at the University of Wisconsin in March 1990. The author would like to thank Eric Arnett, Tad Homer Dixon, Hal Feiveson, Willett Kempton, Michael Lerner, Richard Matthew, Martha Snodgrass, Rob Socolow, Valerie Thomas, Paul Wapner and Wesley Warren for providing valuable comments on earlier drafts of this chapter.

1. For an overview of the role of environmental factors in early theories of politics, see Daniel Deudney, 'Earlier Theories of the Influence of Geography and the Environment Upon Politics', *Global Geopolitics: Materialist World Order Theories of the Late 19th and Early 20th Centuries* (Unpublished PhD dissertation, Politics Department, Princeton University, 1989) Chapter III.
2. For overviews of the global warming problem, see Dean Edwin Abrahamson (ed.), *The Challenge of Global Warming* (Washington, DC: Island Press, 1989).
3. William D. Ruckelshaus, 'Toward a Sustainable World', *Scientific American* (September 1989) p. 167.
4. See, for example, Ernst Callenbach, *Ecotopia* (Berkeley, CA: Banyan Books, 1975); Ernst Callenbach, *Ecotopia Emerging* (Berkeley, CA: Banyan, 1981); John Stadler (ed.), *Eco-Fiction* (New York: Simon and Schuster, 1971); and Whitley Strieber and James Kunetka, *Nature's End: The Consequences of the Twentieth Century* (New York: Warner, 1986).
5. Definitions of the scope of the 'state' vary greatly. At one extreme are those, such as Alan James, who define a state as 'territory, people and a government' in Alan James, *Sovereign Statehood* (London: Allen and Unwin, 1986) p. 13. More narrowly, the state can mean the state administrative apparatus, as when Theda Skocpol defines the state as 'a set of administrative, policy and military organisations

headed, and more or less well coordinated, by an executive authority' in Theda Skocpol, *States and Social Revolutions* (Cambridge: Cambridge University Press, 1979) p. 29. For a penetrating review of these conceptual problems see Fred Halliday, 'State and Society in International Relations: A Second Agenda', *Millennium: Journal of International Studies* (Vol. 16, No. 2, Summer 1987). For a general analysis of the state's complex role in contemporary politics, see John A. Hall and G. John Ikenberry, *The State* (Minneapolis, MI: University of Minnesota Press, 1989).

6. Stanley Hoffmann, 'International Relations: An American Social Science', *Daedalus* (Summer 1977) p. 53.
7. James Rosenau, 'The State in an Era of Cascading Politics: Wavering Concept, Widening Competence, Withering Colossus, or Weathering Change?', in James Caporaso (ed.), *The Elusive State* (Beverly Hills, CA: Sage, 1989).
8. Lester Brown, *Redefining National Security* (Washington, DC: Worldwatch Paper, No. 14, October 1977); Michael Renner, *National Security: The Economic and Environmental Dimensions* (Washington, DC: Worldwatch Paper, No. 89, May 1989); Jessica Tuchman Mathews, 'Redefining Security', *Foreign Affairs* (Vol. 68, No. 2, 1989) pp. 162–77; and Norman Myers, 'Environmental Security', *Foreign Policy* (No. 74, 1989) pp. 23–41.
9. Richard Ullman, 'Redefining Security', *International Security* (Vol. 8, No. 1, Summer 1983) pp. 129–53.
10. Hal Harvey, 'Natural Security', *Nuclear Times* (March/April 1988) pp. 24–6.
11. Philip Shabecoff, 'Senator Urges Military Resources Be Turned to Environmental Battle', *The New York Times* (29 June 1990) p. 1A.
12. 'Strategic Environmental Research Program', *Congressional Record* (28 June 1990) pp. S.8929–43.
13. Quentin Skinner, 'Language and Political Change', and James Farr, 'Understanding Political Change Conceptually', in Terence Ball *et al.* (eds), *Political Innovation and Conceptual Change* (Cambridge: Cambridge University Press, 1989).
14. A particularly lucid and well-rounded discussion of security, the state and violence is found in Barry Buzan, *People, States, and Fear: The National Security Problem in International Relations* (Chapel Hill, NC: The University of North Carolina Press, 1983) particularly pp. 1–93. See also Arnold Wolfers, 'National Security as an Ambiguous Symbol', in *Discord and Collaboration* (Baltimore, MD: Johns Hopkins University Press, 1962).
15. Arthur Westing, *Warfare in a Fragile World: Military Impact on the Human Environment* (London: Taylor & Francis in association with the Stockholm International Peace Research Institute, 1980).
16. Zhores A. Medvedev, *Nuclear Disaster in the Urals* (New York: Norton, 1979).

17. Michael Howard, 'War and the Nation-State', *Daedalus* (Fall 1979). Emphasis in original.
18. See, in particular, Peter Gleick, 'The Implications of Global Climatic Changes for International Security', *Climatic Change* (Vol. 15) pp. 309–25; and Peter Gleick, 'Global Climatic Changes and Geopolitics: Pressures on Developed and Developing Countries', in A. Berger *et al.* (eds), *Climate and Geo-Sciences, 1989* (Amsterdam: Kluwar Academic Publishers, 1989).
19. Arthur H. Westing, 'Global Resources and International Conflict: An Overview', in Arthur H. Westing (ed.), *Global Resources and Environmental Conflict: Environmental Factors in Strategic Policy and Action* (Oxford: Oxford University Press in association with the Stockholm International Peace Research Institute, 1986) p. 1.
20. For a useful survey of theories relevant for such analysis, see Tad Homer Dixon, *Environmental Change and Human Conflict* (Cambridge, MA: Working Paper, American Academy of the Arts and Sciences, 1990).
21. For discussions of resource autarky during the 1930s, see Brooks Emeny, *The Strategy of Raw Materials* (New York: Macmillan, 1934); Norman Rich, *Hitler's War Aims: Ideology, The Nazi State, and the Course of Expansion* (New York: W.W., Norton & Co., 1973); and William Carr, *Arms, Autarchy, and Aggression: A Study in German Foreign Policy, 1933–1939* (London: Edward Arnold, 1972). The reigning assumption of the era is captured in the following passage:

    > . . . capacity for industrialization [is the key to a state's future power position]. And since large-scale industrialization presupposes the possession or ready availability of vast quantities of basic industrial materials, nature, through her unequal distribution of these, has rigidly set the limit to the number of states capable of achieving the status of Great Powers.

    Brooks Emeny, *The Strategy of Raw Materials* (New York: Macmillan, 1934) p. 1.
22. James Crowley, *Japan's Quest for Autonomy: National Security and Foreign Policy, 1930–1938* (Princeton, NJ: Princeton University Press, 1966).
23. Nicholas John Spykman, *America's Strategy in World Politics: The United States and the Balance of Power* (New York: Harcourt, Brace and Co., 1942).
24. Alfred E. Eckes, Jr., *The United States and the Global Struggle for Minerals* (Austin, TX: University of Texas Press, 1979).
25. And more generally, see Richard Rosecrance, *The Rise of the Trading State: Commerce and Conquest in the Modern World* (New York: Basic Books, 1986).

26. Ronnie D. Lipschutz, *When Nations Clash: Raw Materials, Ideology and Foreign Policy* (New York: Ballinger, 1989).
27. Among the most recent versions of the argument that war is of declining viability are: Evan Luard, *The Blunted Sword: The Erosion of Military Power in Modern World Politics* (New York: New Amsterdam Books, 1989); and John Mueller, *Retreat from Doomsday: The Obsolescence of Major War* (New York: Basic Books, 1989).
28. H.E. Goeller and Alvin Weinberg, 'The Age of Substitutability', *Science* (Vol. 201, 20 February 1967). For some recent evidence supporting this hypothesis, see Eric D. Larson, Marc H. Ross and Robert H. Williams, 'Beyond the Era of Materials', *Scientific American* (Vol. 254, 1986) pp. 34–41.
29. For analyses of the vulnerability of energy infrastructure to contemporary means of destruction, see: Wilson Clark and Jake Page, *Energy, Vulnerability, and War* (New York: Norton, 1981); and Amory and Hunter Lovins, *Brittle Power* (Andover, MA.: Brick House, 1982).
30. For discussion of authoritarian and conflictual consequences of environmentally-constrained economies, see: William Ophuls, *Ecology and the Politics of Scarcity* (San Francisco, CA: Freeman, 1976) p. 152. See also Susan M. Leeson, 'Philosophical Implications of the Ecological Crisis: The Authoritarian Challenge to Liberalism', *Polity* (Vol. 11, No. 3, Spring 1979); Ted Gurr, 'On the Political Consequences of Scarcity and Economic Decline', *International Studies Quarterly* (Vol. 29, 1985) pp. 51–75; and Robert Heilbroner, *An Inquiry Into the Human Prospect* (New York: W.W. Norton, 1974).
31. See, for example, Robert Gilpin, *War and Change in World Politics* (Cambridge: Cambridge University Press, 1981).
32. Bernard Brodie, 'The Impact of Technological Change on the International System', in David Sullivan and Martin Sattler (eds), *Change and the Future of the International System* (New York: Columbia University Press, 1972) p. 14.
33. For a particularly lucid argument that the nation-state system is over-developed relative to its actual problem-solving capacities, see George Modelski, *Principles of World Politics* (New York: The Free Press, 1972).

Index

Numbers in bold refer to figures or tables.